W9-AVL-735

BEAMAN MEMORIAL PUB LIBY
8 NEWTON ST
WEST BOYLSTON MA 01583

1994

WITHDRAWN

THE
EARTH
ATLAS

By Susanna van Rose
Illustrated by Richard Bonson

DORLING KINDERSLEY
LONDON • NEW YORK • STUTTGART

A DORLING KINDERSLEY BOOK

Senior Art Editor Martyn Foote
Project Editor Laura Buller
US Editor B. Alison Weir
Production Shelagh Gibson
Managing Editor Susan Peach
Managing Art Editor Jacquie Gulliver
Consultant Keith Lye

A 374090015389098

First American edition, 1994
2 4 6 8 10 9 7 5 3 1

Published in the United States by
Dorling Kindersley Publishing, Inc.
95 Madison Avenue,
New York, NY 10016

Copyright © 1994
Dorling Kindersley Limited, London

All rights reserved under International and Pan-American
Copyright Conventions. No part of this publication may be
reproduced, stored in a retrieval system, or transmitted in any form
or by any means, electronic, mechanical, photocopying, recording,
or otherwise, without the prior written permission of the copyright
owner. Published in Great Britain by Dorling Kindersley Limited.
Distributed by Houghton Mifflin Company, Boston.

Library of Congress Cataloging-in-Publication Data:

Van Rose, Susanna.
The Earth Atlas / written by Susanna van Rose : illustrated by
Richard Bonson. — 1st American ed. 1994.
p. cm.
Includes index.
ISBN 1-56458-626-X
1. Physical geography—Maps for children. 2. Geology—Maps for
children. 3. Children's atlases. [1. Atlases.] I. Bonson, Richard.
ill. II. Title.
G1046.C1V3 1994 <G&M> 94–8765
912—dc20 CIP
MAP AC

Reproduced in Essex by Dot Gradations
Printed in Italy by New Interlitho, Milan

CONTENTS

JUV
912
VAN

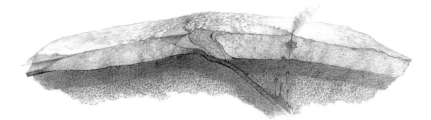

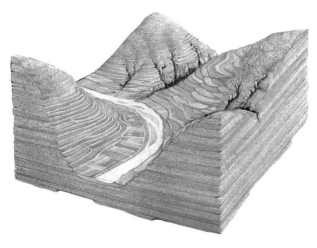

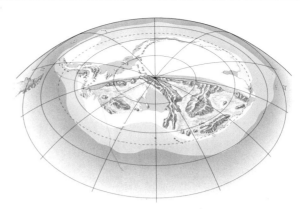

PUTTING THE EARTH IN A BOOK

HOW CAN THE ENTIRE EARTH – from high mountains to deep valleys, from vast oceans to flat deserts – be squeezed between the pages of a book? In order to explain how the Earth looks, this book uses several different ways to show the round Earth on flat paper. The rocky outer surface that humans and animals live on is just a small part of the whole Earth. To understand the forces that have shaped and changed its surface, the illustrations in this book sometimes cut through the Earth's layers right down into its fiery center. In addition, maps, photographs, and diagrams help to show the huge variety of the Earth's landscapes, and explain how they were created. Together, the elements in this book show how the Earth really works.

The Americas *Africa and Europe* *East Asia and Oceania* *The Pacific Rim*

Four faces of the Earth

Because the Earth is round, it is not possible to see all sides at once. Above are four faces, or views, of the globe, each focusing on a different area. One of these four faces appears on most pages of the book, to give a general idea of the location of the main illustration for that page.

A page explained

The double page below shows how information is presented in this book. An introduction gives an outline of the most important facts and ideas. Most pages feature an illustration of a particular place on Earth, chosen to represent a particular geographical feature. Details are explained by smaller illustrations.

Box marks the area shown in the main illustration

Box marks the area shown on the map

Understanding the maps

Each globe view of the Earth includes a box marking the area shown on a more detailed, flat map. A section of the Earth's surface is cut from this map, then lifted out and laid down with the sky uppermost. This section is featured in the main illustration. Areas not shown are marked with a dashed line.

Marked area in the middle is left out of the illustration

Illustration features these two areas from the map

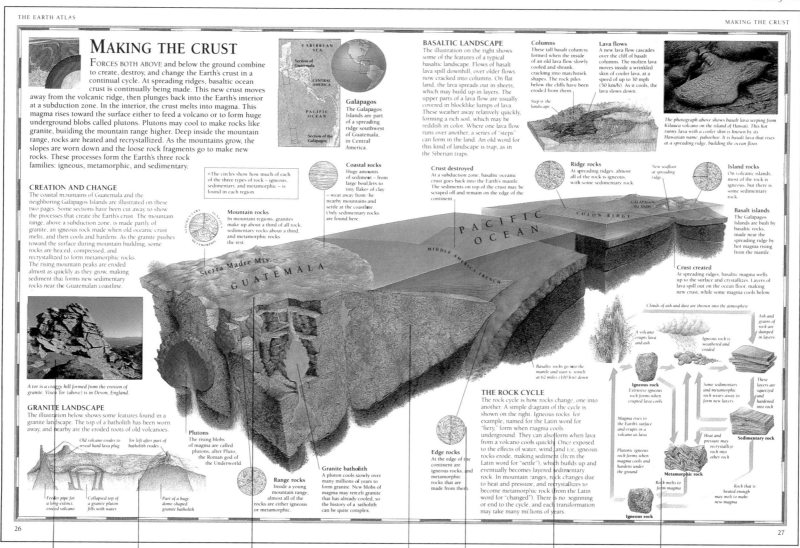

Photographs
These show what landscapes look like when they are made of the rocks, or formed by the geological processes, described on the double-page spread.

Place names help to locate the illustrations

Small illustrations show details of landscapes related to the main illustration

Main illustrations
Different colors are used to show the different kinds of rock that are found in the places featured in the main illustrations.

This detail shows the rock types found in this part of the Earth

Captions
These work closely with the illustrations to present facts and information.

Charts and diagrams
Ideas that relate to the main illustration are often explained by charts and diagrams. Some step-by-step diagrams show how a landscape was formed.

Cutting up mountains

Some of the main illustrations feature huge chunks of the Earth. This one, representing the Himalayas and Tibet. shows a section of the Earth which is really hundreds of miles (kilometers) long. The Himalaya Mountains, the highest mountains on Earth, look tiny on the page.

The section of the Earth featured in the illustration is boxed on the globe view

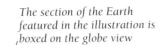

The section is cut out, lifted, and then laid flat

Once the illustration is laid flat, it is easier to understand

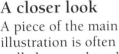

Ancient maps (right) were based as much on imagination as on fact. Today, mapmakers can check the accuracy of their measurements by comparing their maps with satellite photographs (above).

Mapping the Earth

Although the nearly round Earth is most accurately represented as a globe, flat maps provide an almost complete picture of a large area. The maps in this book are used to show specific areas featured in the illustrations. Others show the major mountain ranges, deserts, and frozen regions on Earth.

A closer look

A piece of the main illustration is often pulled out and made bigger, so that details are easier to see. This cross-section, for example, shows what the rocks inside the mountains look like.

This block has been cut out of the main artwork and made larger

Below the surface

The pulled-out blocks of the illustration show us not only the land surface, but also what is going on in the layers underneath. Sometimes this is important in helping to understand the surface landscape and rocks.

These dashed line areas show sections that are left out

This is the most distant section

This is a section taken from the middle of the glacier

This is the end, or snout section of the glacier

Cutouts

This illustration of the Athabasca Glacier in Canada shows inside and underneath the glacier – views it is not usually possible to see. The inside of the glacier is shown in cross-sections of, or cuts through, the illustration, which are pulled apart to give a better view. In one section of the illustration the whole glacier is removed, along with all the boulders and rock fragments trapped in the ice, to show the rocky surface below.

Times change

Landscape is always changing, but usually very slowly. This book explains the forces that are continually changing the Earth's surface. Sometimes change happens rapidly – the volcanic eruption at Mount St. Helens, Washington, blew the side of the mountain away in only hours (right). Other changes have happened slowly, over tens of thousands of years, or even over a time span long before humans had evolved. Illustrations such as the sequence below are used to show these changes.

This is Mount St. Helens before its eruption.

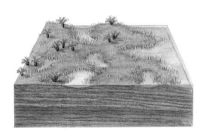

This is a landscape as it might have looked 100 million years ago. Dinosaurs roamed over swamps, and left their footprints in the wet sand.

This is the same place 40 million years later. The dinosaurs are now extinct, their footprints long since buried. The land has sunk and is under a shallow, young sea.

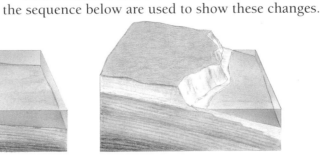

Today the sea is gone, and dry land is left behind. The muddy ooze that was on the seabed has hardened into chalk, which is being eroded into steep cliffs.

After the eruption, there is a large hole where the mountain summit once was.

EARTH IS UNIQUE

AMONG THE PLANETS of the solar system, Earth is unique in many ways. It is enveloped in a cloud of gas called the atmosphere. One gas in particular, oxygen, makes a very special contribution: supporting life. Oxygen makes up a critical one-fifth of the Earth's total atmosphere; if there were slightly more oxygen in the atmosphere, plants would spontaneously catch fire. Water covers most of the Earth's surface, keeping temperatures moderate and releasing essential water vapor into the atmosphere. Inside the Earth, restless currents of molten rock create a magnetic field which shields the Earth from space radiation. Other internal movements cause the surface to slowly churn over, so that the crust constantly renews or reshapes itself.

Green planet
From space, the plants on Earth's land surface make the continents look green. Plants take in carbon dioxide gas from the atmosphere and use it to grow new leaves and branches. In the process they give out oxygen, used by animals and people.

Land planet
Earth's land areas cover only a small part of its surface, where the continents rise above sea level.

THE INNER EARTH
Nothing seems as substantial as the rocky surface we live on, but in relation to the vast inner layers of the Earth it is no more than an eggshell. The surface rocks are made mostly of the chemical elements oxygen and silicon, along with some metals. Beneath this is a thick layer of heavier rock called the mantle, which encloses the inner and outer cores. The solid metal inner core is the densest part of the Earth. The liquid outer core is continually moving, causing a constantly changing magnetic field to envelop the planet.

Air planet
The mixture of gases and water vapor wrapped around the Earth is its atmosphere. Its swirling white clouds are continually on the move, warmed or cooled by the land and sea below, or whipped around by the Earth's rotation.

Water planet
Three-quarters of the Earth's surface is covered by water, most of it within its vast oceans.

Crust
The solid, rocky surface that we live on is called the crust. It is thickest under the continents, and thinnest below the oceans.

Lithosphere and asthenosphere
The outermost 62 miles (100 km) of the mantle is firmly attached to the Earth's outer crust. Together, these make up the lithosphere (from the Greek word for "rock sphere"). Below the lithosphere is a layer of hotter, softer rock another 62 miles (100 km) thick. This is called the asthenosphere, from the Greek word for "weak sphere."

Earth imagined
Ancient people had no way of knowing about the inner Earth. The painting below illustrates the world system of the Babylonians. The Earth is a round, hollow mountain resting on water. The sky, with fixed stars, meteors, and planets, forms a hollow chamber. The sun rises each day through a door in the east, and sets through a door in the west. Even today, no one has ever seen or taken a sample of the Earth's mantle and core. Most of what we know is based on indirect measurement.

Mantle
Making up about nine-tenths of the total bulk of the Earth, the mantle is a thick shell of hot, rocky silicate (silicon and oxygen) minerals. Although the mantle is solid, it does slowly circulate over millions of years.

Outer core
Surrounding the solid inner core is the liquid outer core. Its molten iron and nickel metals are at the slightly cooler temperature of 10,000°F (5,500°C), and are under less pressure than the inner core. The movement of the hot liquid here generates the Earth's magnetic field. Changes in the strength and the polarity of the magnetic field are linked to changes in the circulation.

Inner core
At the heart of the planet is its inner core, made of the same metals that make up the outer core. The temperature is so high here – up to 10,800°F (6,000°C) – that these metals should be molten liquid. But the immense pressure is enough to compress them into solids.

Summer solstice

On June 21, the tilt of the Earth's axis means the strongest and most direct light from the sun is in the northern hemisphere. The longest day of the year occurs at this time; the Arctic Circle has 24 hours of daylight.

Northern hemisphere summer

Note: This diagram is not to scale.

Northern hemisphere spring

Spring equinox

At the spring equinox on March 21, the sun is directly overhead at the Equator. Days and nights have equal length all over the Earth. Equinox means "equal nights."

Northern hemisphere winter

Autumnal equinox

As in the spring equinox, the sun is directly overhead at the equator during the autumnal equinox. Day and night are once again the same length of time in both hemispheres.

Northern hemisphere autumn

Winter solstice

On December 21, the tilt of the Earth means the sun's direct light and heat strike south of the Equator. This is the longest day of the year in the southern hemisphere. The northern hemisphere has its shortest day and longest night.

THE EARTH'S ORBIT

Like the other planets, Earth travels around the sun; it takes a year to make one complete round-trip. The pathway is called its orbit. Earth's orbit is not quite circular, but is slightly oval in shape. As it makes its orbit, Earth is also spinning around an imaginary line running between the north and south poles, called its axis. Each complete spin takes about 24 hours. When light from the sun illuminates half the globe, this region has day while the dark half has night. Our seasons (above) are also a result of the Earth's orbit and rotation.

A beautiful result of the Earth's magnetism is an aurora – a shimmering display of light seen in the night sky in and near Arctic and Antarctic regions. The Earth's magnetic poles attract charged particles in the atmosphere which radiate colored light.

The Earth's plates

The Earth's lithosphere is broken into fewer than a dozen large and many smaller plates. These move slowly and steadily. Everything carried on them moves, too – from huge continents to entire oceans.

Hydrosphere

The watery layer of the Earth, the hydrosphere, includes its oceans, lakes, rivers, underground water, and snow and ice. The hydrosphere covers the oceanic crust almost totally, to an average depth of 3 miles (5 km). The water laps onto and submerges the edges of the continents.

Atmosphere

The total thickness of the atmosphere is at least 620 miles (1,000 km), with almost all of the gas molecules hugging tight to the Earth. The atmosphere is densest in the lowest 6 miles (10 km), the troposphere. Above this is the stratosphere, where the atmosphere is thinner because it contains fewer molecules.

A gas within the atmosphere called ozone shields the Earth from some of the sun's rays

Upper part of the mantle and the crust forms the lithosphere

Thick continental crust

Thin oceanic crust

Hydrosphere

Troposphere

Stratosphere

Mesosphere

Thermosphere

Exosphere and magnetosphere

THE OUTER EARTH

The solid rocky surface of the Earth is a place of continual change. Here the Earth's rocks meet – and interact – with the watery layer of the hydrosphere and the gassy layer of the atmosphere. The rocks react both chemically and physically with these turbulent layers. Moving air and water, and temperature changes, break down the rocks physically. Oxygen from the atmosphere reacts with silicates in the rocks to change them chemically, making new rocky minerals. The Earth's outer layers are so closely interlinked that a change in one layer is likely to affect the others.

BOMBARDMENT FROM SPACE

THE EARLY HISTORY OF THE EARTH is shrouded in mystery. Most clues about the young planet have long been worn away. But the events of this turbulent first billion years – nearly a quarter of the Earth's history – set the scene for the planet of today. During that time, the Earth evolved out of a cloud of dust to a cooling, encrusted planet wrapped in an envelope of gases. Its first rocky skin was ripped apart by the impact of huge meteorites. This bombardment of the Earth and other planets went on while the crust grew thick enough to withstand the impact, and cool enough for rainwater to collect in pools – the earliest oceans.

Chemical clouds
Billowing clouds of steam and smelly chemical gases poured from the Earth. Gravity held most of these near the surface, but some of the lighter gases escaped.

Crust cooling over hot interior

Cooler crust
The thin surface crust over the Earth gradually grew thicker. From time to time, slabs of cooled crust probably plunged back into the molten mantle below and were melted again. Wherever the crust was cooling and crystallizing, thick clouds of gases bubbled out, building the Earth's first atmosphere.

Meteorites pound the Earth
One after another, meteorites pounded the molten surface of the Earth. A thin, brittle skin barely covered the Earth; it was not yet tough enough to resist the impact of the falling meteorites.

Piercing the skin
As a meteorite landed, hot liquid rock splashed out around the hole in the surface. The meteorite itself plunged into the hot interior. Torrents of lava probably surged up to the hole punched in the surface, spilling out around it in thick sheets.

Large meteorite about to land on the Earth

Flows of lava spreading from the impact crater

Meteorites
These arrivals from space are pieces of an exploded planet, one of the Earth's long-ago neighbors. Some, like this meteorite, are rich in iron, and probably came from the core of the exploded planet.

Craters
Meteorites make an impact crater when they land. The impact may be so intense that the meteorite shatters or even evaporates altogether. Meteor Crater in Arizona is 3,936 ft (1,200 m) across and 557 ft (170 m) deep.

Lava fountains
Lazy molten lava bubbles underneath a huge volcano. When it erupts, gas escapes from the hot lava, and its intense bubbling forces the lava to spill out in fountains.

Dust from volcanic eruptions can cause lightning because it is electrically charged by friction

The first rainfall
The steam escaping from the cooling crust may condense to form rain clouds. If the clouds become big or cool enough, rain falls.

The surface is still so hot that rain instantly turns to steam

Lava layers
Massive volcanoes pour out seething hot lava, which spreads out in runny sheets to cover the surface. Layer by layer, the lava builds the volcanoes higher, and thickens the cooling crust.

Hot water collects in steaming pools on top of the hardening layers of lava

Sizzling surface
At first, rain that falls to the Earth sizzles like water on a hot griddle. But after many rainstorms, the surface of the Earth is cool enough for pools of steaming water to collect.

Sizzling water boils as it skids over the hot rocky surface.

HOT FIERY PLANET
The Earth was probably a hot, fiery planet around 4.5 billion years ago, covered with a thin, rocky skin. The skin cracked and remelted over and over, solid crusts sinking back into the hot interior to melt again, while new crust grew and cooled. Like the other planets in the solar system, the Earth was continually bombarded by meteorites – loose rock fragments plummeting onto its thin skin. These may have come from the explosion of an older planet that had a rocky surface and an iron-rich core. The asteroids orbiting Earth today may be the remains of this exploded planet.

THE THICKENING CRUST
As the molten Earth began to grow a thicker, more solid crust, the cooling lava at the surface bubbled and gave off vast clouds of gases. These became the Earth's first atmosphere, a smelly mixture very different from the atmosphere today. It included steam, carbon dioxide, and nitrogen – but no oxygen. Some of its gases, such as the lightweight hydrogen and helium, easily escaped Earth's gravity. Others, such as hydrogen chloride, became combined with rocks during weathering.

THE FIRST OCEANS
Volcanoes punctuated the Earth's still hot, steamy surface. The gases that poured out of the volcanoes with each eruption were added to the thick, choking chemicals of the atmosphere. Some volcanic mountains may have grown high enough to cause the water in the atmosphere to cool and condense (turn back into liquid). The tiny drops of water gathered to form clouds, so that it could rain. At first, the solid surface of the Earth may have been so hot that the rain evaporated as soon as it fell. Eventually, some parts of the surface became cool enough for the rain to collect there in hot pools.

Ropelike coils are found today in a type of lava called "pahoehoe"

Bubbling pool
A bubbling hot pool of sulfurous water has some mud at the bottom, from the first weathering of rocks.

The steam rises and may condense again to make clouds

Cooling coils
As the skin of the lava cools, it is twisted and pulled by the movements of the still-molten lava into ropelike coils.

Rock reactions
Rocks reacted chemically with the gases in the primitive atmosphere – the first signs of chemical weathering, the breaking down of the Earth's rocky materials. For example, sulfur gases and hydrogen chloride react with minerals in the rocks to transform them into clay minerals, forming the first muds on Earth.

DAWN OF EARTH HISTORY

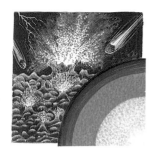

THE ARCHEOZOIC HISTORY of the Earth, from the Greek word for "beginning," represents its first 2 billion years. After its rocky crust formed and the surface cooled enough for steaming pools of rainwater to collect, the Earth entered a new stage of development. During this dawn of Earth history, the continents came into being. They were made of granite-like rocks that thickened the crust and eventually built the first mountains. The oceans were not just low places filled with water; instead, they had a rocky crust quite different from that of the continents. How this separation into two types of crust happened, we do not know. Oxygen was added to the atmosphere from tiny sea plants. Although much of the oxygen was at first consumed by chemical reactions, in time, enough built up to support primitive life forms. Then, 600 million years ago, there was an explosion of life forms with skeletons. At last, the stage was set for the Earth we know today.

Hot lava
Komatiite lava erupted at temperatures as high as 3,100° F (1,700° C) – twice as hot as many lavas today.

THE ARCHEOZOIC EARTH
It is not easy to reconstruct the Archeozoic Earth, because the rocks of that age have changed so much since they formed. Around 3.5 billion years ago the continents began to form as light, molten rock similar to granite rose under giant volcanoes. The volcanoes have long ago been eroded away, but their granite "roots" formed the first continents. An unusual lava unlike those known today formed the first ocean crust. This kind of lava, called komatiite, seems to have been very runny and very hot – so hot that it melted the rocks it flowed over.

Fast flowing
Some of the fastest-flowing lavas today move at about 30 mph (50 km/h), but runny komatiite moved much more quickly.

Komatiite crystals
Komatiite was very much hotter than modern lavas. It cooled to form huge crystals, which looked like coarse blades of grass.

Cool crust
This lava has cooled to form a rough-textured crust of rock.

Traces in the rocks
Today, the only traces of the first ocean floor are tiny rock scraps containing komatiite. It seems that whatever mechanism made this strange lava stopped happening 2.5 billion years ago. The modern ocean crust is made from a very different hardened lava known as basalt.

Slow the flow
As a lava flow cools, it becomes stickier and more solid, and it slows down.

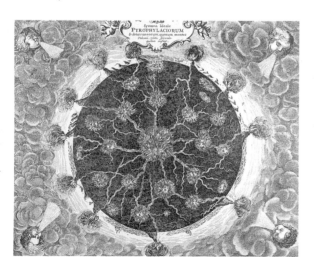

This 17th-century engraving by Anthanasius Kircher illustrates his theory that the Earth's interior was still molten, and that perpetual fire fed volcanoes. His theory was not accurate for the Earth of today, but could have described its first billion years.

A river of lava
When komatiite lava erupted, it melted the rocks over which it flowed. As a result, it cut its path or channel lower and lower, just like a river.

Crusty black cooled lava

Layers of old cooled lava

Cutting through the rocks
A komatiite lava flow that flowed for a week could cut a channel 65 ft (20 m) deep, just by melting the rocks underneath.

Quick-moving lava shooting through the crusty channel

Long-distance lava
The lava is able to flow through its eroded channel for long distances from its original eruption.

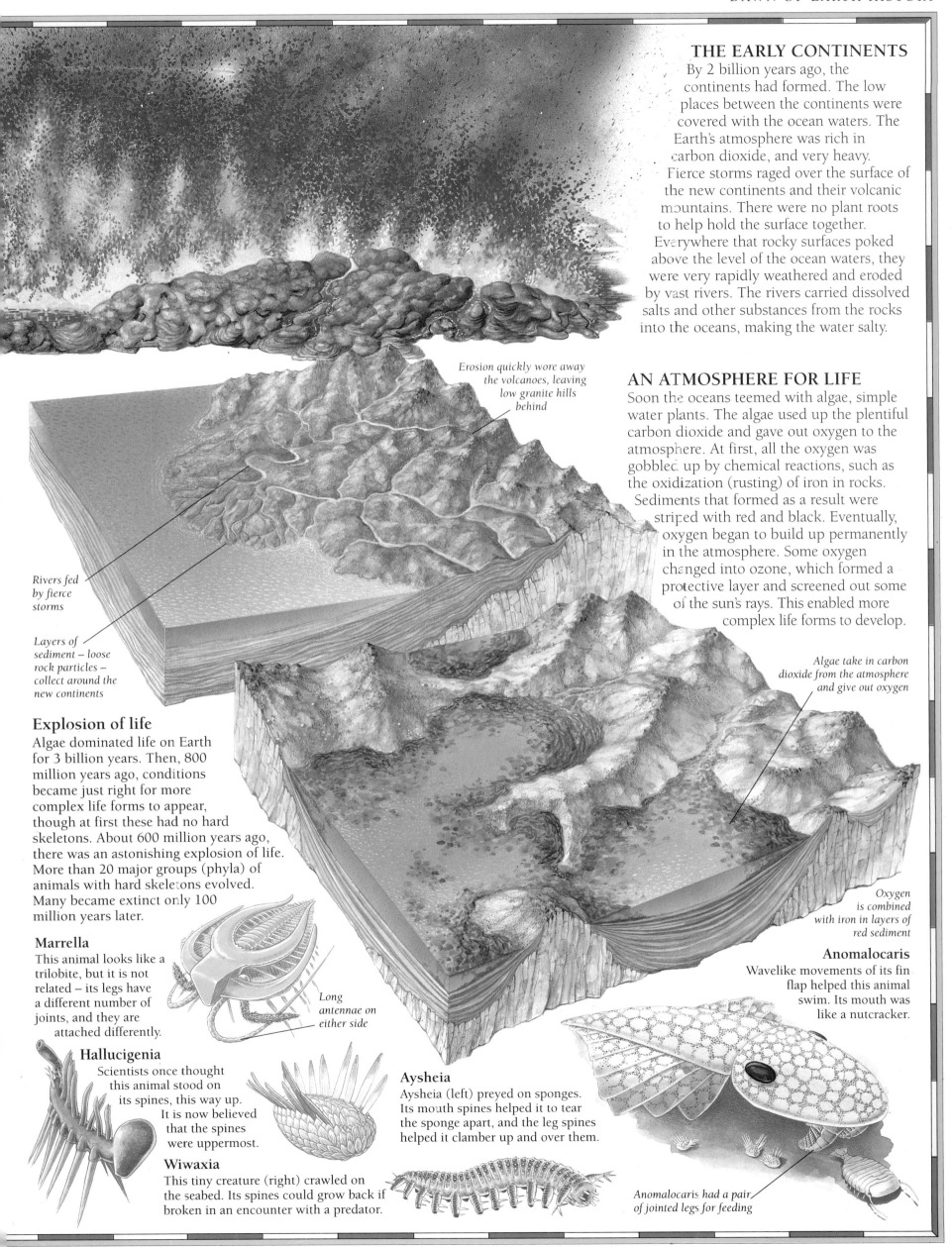

THE EARLY CONTINENTS

By 2 billion years ago, the continents had formed. The low places between the continents were covered with the ocean waters. The Earth's atmosphere was rich in carbon dioxide, and very heavy. Fierce storms raged over the surface of the new continents and their volcanic mountains. There were no plant roots to help hold the surface together. Everywhere that rocky surfaces poked above the level of the ocean waters, they were very rapidly weathered and eroded by vast rivers. The rivers carried dissolved salts and other substances from the rocks into the oceans, making the water salty.

AN ATMOSPHERE FOR LIFE

Soon the oceans teemed with algae, simple water plants. The algae used up the plentiful carbon dioxide and gave out oxygen to the atmosphere. At first, all the oxygen was gobbled up by chemical reactions, such as the oxidization (rusting) of iron in rocks. Sediments that formed as a result were striped with red and black. Eventually, oxygen began to build up permanently in the atmosphere. Some oxygen changed into ozone, which formed a protective layer and screened out some of the sun's rays. This enabled more complex life forms to develop.

Erosion quickly wore away the volcanoes, leaving low granite hills behind

Rivers fed by fierce storms

Layers of sediment – loose rock particles – collect around the new continents

Algae take in carbon dioxide from the atmosphere and give out oxygen

Oxygen is combined with iron in layers of red sediment

Explosion of life

Algae dominated life on Earth for 3 billion years. Then, 800 million years ago, conditions became just right for more complex life forms to appear, though at first these had no hard skeletons. About 600 million years ago, there was an astonishing explosion of life. More than 20 major groups (phyla) of animals with hard skeletons evolved. Many became extinct only 100 million years later.

Marrella

This animal looks like a trilobite, but it is not related – its legs have a different number of joints, and they are attached differently.

Long antennae on either side

Hallucigenia

Scientists once thought this animal stood on its spines, this way up. It is now believed that the spines were uppermost.

Wiwaxia

This tiny creature (right) crawled on the seabed. Its spines could grow back if broken in an encounter with a predator.

Aysheia

Aysheia (left) preyed on sponges. Its mouth spines helped it to tear the sponge apart, and the leg spines helped it clamber up and over them.

Anomalocaris

Wavelike movements of its fin flap helped this animal swim. Its mouth was like a nutcracker.

Anomalocaris had a pair of jointed legs for feeding

LINES OF FIRE

MOST VOLCANOES lie along lines of fire that encircle the planet. These are places where the turbulence of the Earth's fiercely hot interior is revealed along giant cracks at the surface. Many earthquakes also happen near these cracks. This uneven distribution on the Earth was noticed early in the nineteenth century, though at that time there was no way of explaining why it should be. An especially large number of volcanoes and earthquakes happen around the shores of the Pacific Ocean, shown here. This explosive area is sometimes called the Pacific Ring of Fire. Its amazing landscapes are a result of its fiery past.

The Valley of Ten Thousand Smokes
A huge ashy eruption blanketed this valley in Alaska in 1912. The first explorers to venture into the valley saw countless bubbling volcanic "smokes" rising from the ash. The ash itself is up to 164 ft (50 m) deep in places.

Key to map
Three types of margins, or boundaries between plates, are shown on this map: constructive margins are fiery red, destructive margins are brown, and conservative margins are purple. They will be explained on the next two pages.

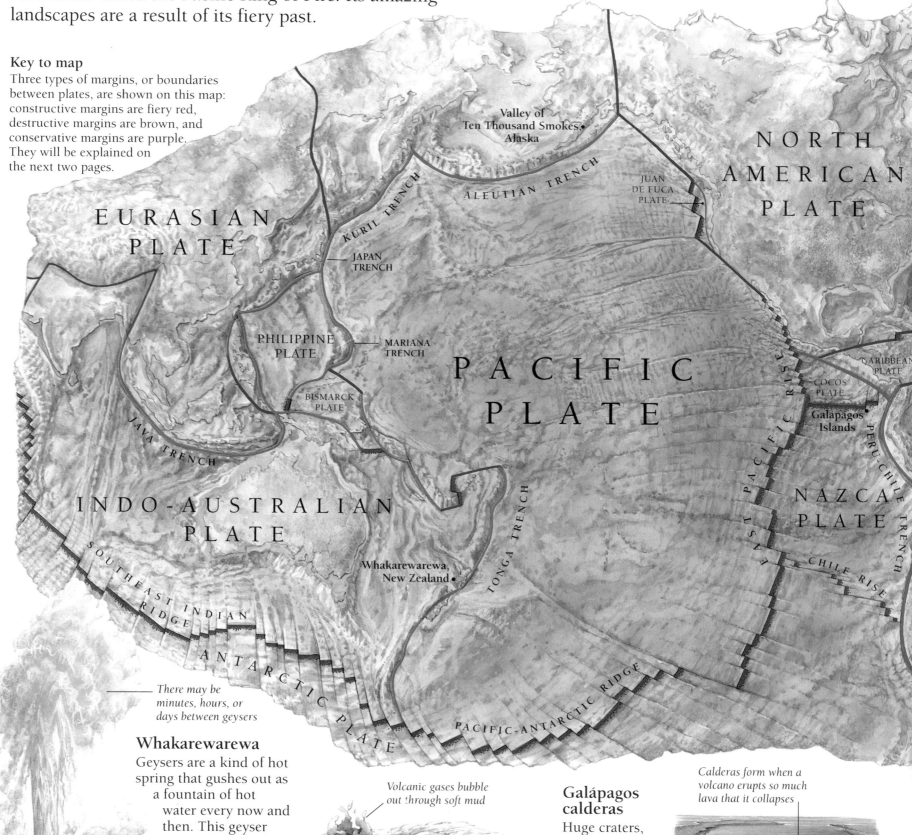

NORTH AMERICAN PLATE

Valley of Ten Thousand Smokes, Alaska

ALEUTIAN TRENCH

JUAN DE FUCA PLATE

EURASIAN PLATE

KURIL TRENCH

JAPAN TRENCH

PHILIPPINE PLATE

MARIANA TRENCH

BISMARCK PLATE

PACIFIC PLATE

PACIFIC RISE

CARIBBEAN PLATE

COCOS PLATE

Galápagos Islands

PERU-CHILE TRENCH

JAVA TRENCH

INDO-AUSTRALIAN PLATE

EAST PACIFIC RISE

NAZCA PLATE

CHILE RISE

TONGA TRENCH

Whakarewarewa, New Zealand

SOUTHEAST INDIAN RIDGE

ANTARCTIC PLATE

PACIFIC-ANTARCTIC RIDGE

There may be minutes, hours, or days between geysers

Whakarewarewa
Geysers are a kind of hot spring that gushes out as a fountain of hot water every now and then. This geyser and these bubbling mud pools are in New Zealand's Whakarewarewa, south of Rotorua.

Volcanic gases bubble out through soft mud

Galápagos calderas
Huge craters, called calderas, pockmark the summits of the Galápagos Islands.

Calderas form when a volcano erupts so much lava that it collapses

Surtsey born from the sea

In 1963 a bubbling volcano broke through the waves south of Iceland. The new island, named Surtsey after an ancient Icelandic fire god, grew larger as more and more lava poured out over the first loose ash layers. Surtsey is now home to a variety of plants, insects, and birds.

Surtsey sits astride the Mid-Atlantic Ridge, where two of the Earth's plates are slowly pulling apart. New magma from inside the Earth cools to heal the crack, making new ocean crust.

A YEAR IN THE LIFE OF A PLANET

About 30 volcanoes are erupting in any one year. Of these, some continue to erupt for several years, or even decades, while others may erupt only once during that time. One or two have been active for thousands of years. Thousands of earthquakes happen each year, but most are far too small to be noticed. A few dozen cause shaking which can be felt, and fewer than ten are really large. There is no sign that today earthquakes and volcanic eruptions are happening either more or less often.

Fingal's Cave

The basalt columns that built the island of Staffa off western Scotland are formed from lava which cracked into regular shapes as it cooled slowly, 60 million years ago.

Le Puy in France

This church in southern France is built atop a volcanic rock 250 ft (76 m) high. The rock hardened inside the volcano two million years ago. It was exposed when softer, ashy rocks were eroded away from around it.

Urgup Cones, Turkey

These pillars in Turkey are sculpted out of volcanic ash, from eruptions of long ago. Volcanic gases, like the "smokes" in Alaska, welded the ash in places. Wind and rain have shaped it into pillars.

Karum salt pillars

Rain weathers salt out from volcanic rocks, washing it into the Assale Lake in Ethiopia. The water is so salty that the surface crystallizes over, with salt pillars growing 10 ft (3 m) overnight.

Map labels: REYKJANES RIDGE · EURASIAN PLATE · Surtsey, Iceland · Fingal's Cave, Scotland · Le Puy, France · Urgup Cones, Turkey · ANATOLIAN PLATE · IRANIAN PLATE · ARABIAN PLATE · Karum Pillars, Ethiopia · AFRICAN PLATE · MID ATLANTIC RIDGE · SOUTH AMERICAN PLATE · RIDGE · Réunion · ATLANTIC-INDIAN RIDGE · SCOTIA PLATE · ANTARCTIC PLATE

Réunion

One of the Earth's largest volcanoes is Réunion Island, which rises from the deep ocean floor to its summit craters 10,068 ft (3,069 m) above sea level.

SIGNS OF THE TIMES

Old, eroded volcanic features in the landscape are an indication that millions or even tens of millions of years ago, there were volcanoes at this place. Not all of these features lie along the lines of fire we know today. This means the old lines were in different places, and shows that the Earth's plates move as time passes. This movement is evidence of change in the churning motions of the Earth's turbulent interior.

THE MOVING CRUST

NOT ONLY IS OUR PLANET spinning through space, but also the Earth's surface is heaving about, though very slowly. Each year, the continents move nearly half an inch (a centimeter) or so – some getting closer together, others moving apart. This sounds like a small amount, but over a million years it adds up to about 6 miles (10 km). The movement happens because the inside of the planet is hot and turbulent. Its motion disturbs the cool rocky surface and causes the huge plates of the crust to move around. New ocean floor is made at spreading ridges; over tens or hundreds of millions of years it moves toward a subduction zone, where it is destroyed. This slow movement of the Earth's plates has been going on for billions of years.

In 1919, German scientist Alfred Wegener (above) proposed the idea that the continents move. His theory is known as continental drift.

THE PLATE DEBATE

Wegener himself wasn't sure how continents moved, imagining them to somehow plow their way through the rocky ocean floor. Wegener's theory has been transformed by the understanding that each plate of the Earth's crust (shown lifted off the mantle in the artwork below) carries both continents and oceans, which move together.

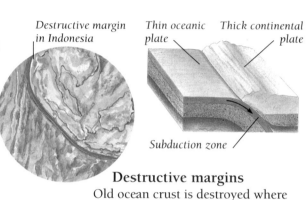

Destructive margin in Indonesia

Thin oceanic plate   *Thick continental plate*

Subduction zone

Destructive margins

Old ocean crust is destroyed where it dives (or subducts) beneath a continent and melts into the mantle. This kind of margin is also called a subduction zone.

PLATE BOUNDARIES

A plate edge meets another at three possible kinds of margins. If a plate carrying an ocean meets a continental plate, the ocean crust plunges down, or subducts, under the continent and disappears. When solid rocky crust of one plate crunches sideways against another equally solid plate, the rocks fracture and earthquakes occur. Where two plates move apart, there is a widening crack in the Earth's outer skin, which fills with hot magma rising from the mantle.

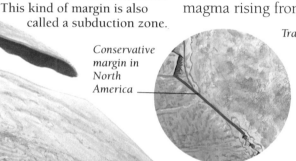

Conservative margin in North America

Transform fault

One plate slides past the other

Conservative margins

Where two plates slide past each other, the rocks grind and crunch and make earthquakes. No new crust is made and no old crust is destroyed. The line along which the plates slide is a transform fault.

Constructive margin in the Atlantic Ocean

Magma oozing up from the mantle

Section of the mantle cut away to reveal the hot rocky material circulating beneath the plates

Earth's turbulent mantle

The mantle is more or less solid, but over long periods of time – millions, or tens of millions of years – it flows like a thick, sticky plastic. Some parts of the mantle are cooler and more solid than others. Other regions, under constructive margins and volcanoes, are hotter and contain some liquid, which will rise to become magma.

Constructive margins

New ocean crust is made where plates are spreading apart. The gap between the plates fills with magma. These margins are also known as spreading ridges.

DRIFTING CONTINENTS

Continual shifting and drifting of the plates which cover the Earth's surface is what changes the shapes of the continents and oceans. Three stages in one continent's drift toward another are shown here. Old ocean crust is swallowed and destroyed at a subduction zone. At the same time, a new ocean grows at the far margin of the plate. Eventually the entire ocean crust disappears back into the mantle, bringing the two continents together. Continental crust cannot subduct – it is too light in weight to go down into the mantle. Instead it grafts onto the other continent.

Ocean crust meets this continent and subducts

Plates pushed apart to make room for the new crust

First push

A spreading ridge pushes one continent toward another. The ocean crust between them has nowhere to go, and begins to subduct at the edge of the far away continent.

Spreading ridge

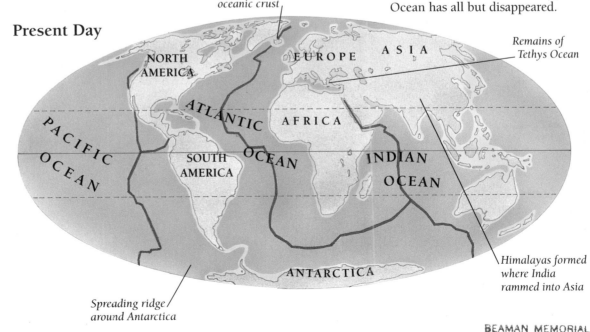

Spreading ridges are almost all under ocean water, except in Iceland (above), where magma is coming to the surface so fast that the ridge has built up a huge land mass made of oceanic crust.

The ocean floor spreads as magma rises, trying to plug the gap between plates, only to be added to the plate edges

Subduction zone

The push continues

As the new ocean grows bigger, the old ocean gets smaller. Some of the old ocean crust is melted at the subduction zone. It rises to feed volcanoes at the surface, and a mountain range starts to grow.

Continental drift means the ocean floor is always renewed – today, there is no ocean crust left on Earth older than 200 million years.

200 Million Years Ago

Enormous Panthalassa (from the Greek for "all seas") Ocean

EURASIA

PANGAEA

Tethys Ocean

Continent of Pangaea (from the Greek for "all lands")

The continents meet

The mountains buckle up and fold as the continents come closer. Finally, there is no more old ocean crust left to subduct. The continents meet, and one grafts onto another, making a bigger continent.

200 million years ago

An ocean separated Europe and Asia (Eurasia) from the southern supercontinent, Pangaea. There was no Atlantic Ocean, and India was attached to Antarctica.

100 Million Years Ago

100 million years ago

Africa and South America had already begun to separate from each other as they broke away from Antarctica. As new ocean floor was built by a spreading ridge between North America and Europe (Laurasia) and Africa, the young Atlantic Ocean was born.

Shrinking Panthalassa Ocean

LAURASIA

SOUTH AMERICA

AFRICA

Tethys Ocean

Atlantic Ocean

Present day

Antarctica is now separated from all the other continents; it is totally surrounded by new ocean floor which has been made by a spreading ridge encircling it. India has rammed into Asia, and the ancient Tethys Ocean has all but disappeared.

Iceland is made of oceanic crust

THE WORLD LONG AGO

The shapes and locations of the continents and oceans were very different in times gone by. It is easy to put together the history of the last two hundred million years, by imagining that today's ocean crust is no longer there. In this way we know that two hundred million years ago there was a giant continent, called Pangaea. This supercontinent came about through the collision and grafting together of yet older continents. It is not so easy to reconstruct what happened before Pangaea. There was an even earlier supercontinent, also pieced together from even older continents which themselves had been formed by previous continental splitting.

Present Day

NORTH AMERICA

EUROPE

ASIA

Remains of Tethys Ocean

ATLANTIC OCEAN

PACIFIC OCEAN

AFRICA

SOUTH AMERICA

INDIAN OCEAN

ANTARCTICA

Spreading ridge around Antarctica

Himalayas formed where India rammed into Asia

BEAMAN MEMORIAL
PUBLIC LIBRARY
WEST BOYLSTON, MASS. 01583

EXPLOSIVE VOLCANOES

EARLY IN 1980, THE SIDE of Mount St. Helens started to bulge. As pressure built up inside the volcano, the bulge got bigger and bigger. On May 18, the entire volcano exploded. Explosive volcanoes like Mount St. Helens (shown here) produce thick, sticky lava, and erupt infrequently. Between eruptions, there is time for gas to build up in the magma below the volcano. Eventually the pressure of the gas blows the overlying rocks apart. The gassy lava froths up and explodes, shattering into tiny fragments. Propelled by the force of the explosion and the continuing release of gas, the fragmented lava billows out. It launches down the steep slope of the volcano as a high-speed ash flow, engulfing everything in its path.

A cloud of ash
The dense cloud of ash from a large eruption may rise 19 miles (30 km) or more into the atmosphere. Winds can carry the choking cloud over great distances. Ash from Mount St. Helens spread more than 150 miles (240 km) in just two hours.

Hot ash
The temperature inside the cloud reached 680°F (315°C).

Blast force
The force of the eruption blew out the side of the mountain.

FIRE MOUNTAIN
Mount St. Helens is shown on the right, with two sections pulled away to expose its explosive center. The last major eruption had been in the 1800s, so the volcano had been silent for over a hundred years before its devastating eruption in 1980. Mount St. Helens at last lived up to the name given to it by its native peoples – Tahonelatchah, or "fire mountain."

Eruption cloud

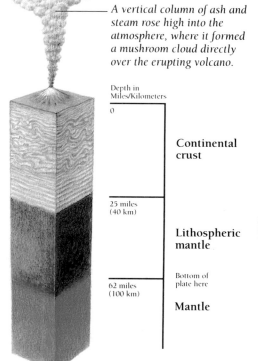

A vertical column of ash and steam rose high into the atmosphere, where it formed a mushroom cloud directly over the erupting volcano.

Depth in Miles/Kilometers

0

Continental crust

25 miles (40 km)

Lithospheric mantle

Bottom of plate here

62 miles (100 km)

Mantle

Mud flows
Glaciers on the volcano slopes were suddenly melted in the eruption, making destructive mud flows which spread far down the river valleys.

Section of mountainside removed to show the interior of the volcano

MAGMA
Magma is generated about 62 miles (100 km) under the ground. It rises through the solid rock in red-hot blobs, and collects in a reservoir. During decades or even centuries of slow cooling, the magma crystallizes and gas bubbles rise to the top. When pressure is suddenly released in an eruption, pressure from the gas blows out the crystals and explodes the magma, which chills to lava glass and falls as ash.

Below the volcano
Lava crystallizes slowly in the magma reservoir below the volcano. The arrival of a new blob of magma from greater depth below may trigger an eruption.

Bubbling under
As the magma rises, the gases contained in it escape and form larger and larger bubbles. These balloon out against the rock, trying to find an escape route.

Washington's Cascade Range

Mount St. Helens is a volcano in the Cascade mountain range, along the northwestern coast of North America.

Bird's-eye view

This color-enhanced photograph of Augustine volcano in Alaska shows the explosive force of an ashy eruption. The plume of ash rises high into the atmosphere, then rains down to cover land and sea. Traces of ash clouds can spread around the Earth, affecting the weather and creating spectacular sunrises and sunsets on the other side of the globe.

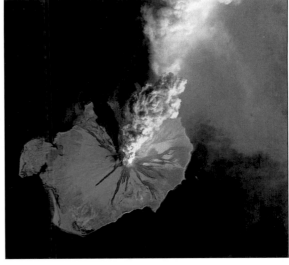

The photograph of the eruption of Augustine volcano was taken from the safety of a Landsat satellite. The ash cloud is about 7 miles (11 km) high.

Avalanche

After a small earthquake shook the volcano, its entire north face trembled slightly, then suddenly broke loose to slide downhill as a massive avalanche.

Rock glows red-hot

Lava bombs

Blocks of hot rock and new magma were catapulted out of the explosion cloud, traveling even faster than the ash cloud itself. Fragments of new lava are known as pyroclasts, meaning fiery fragments. Many are tiny, but together they make up devastating ash flows, which sear every living thing they meet. Ash flows and the eruption clouds flash with lightning bolts, and may cause torrential rains.

Heat blast

The intense heat from the blast scorched the surrounding forest. Some witnesses said that nearby streams became so hot that fish leaped out of them to escape.

THE AFTERMATH

After Mount St. Helens' great ash flow eruptions in 1980, it began to emit thick, almost solid lava, which squeezes slowly up into the crater. It moves by slipping along many tiny cracks that riddle the sticky lava. As this lava builds up inside the crater, it forms a dome shape. Occasionally the dome becomes so steep that hot, solid lava bursts off as an avalanche from the side of the dome. Mount St. Helens will regrow, but it may take tens of thousands of years to rebuild to its original size.

Smaller particles break off as the rock flies through the air

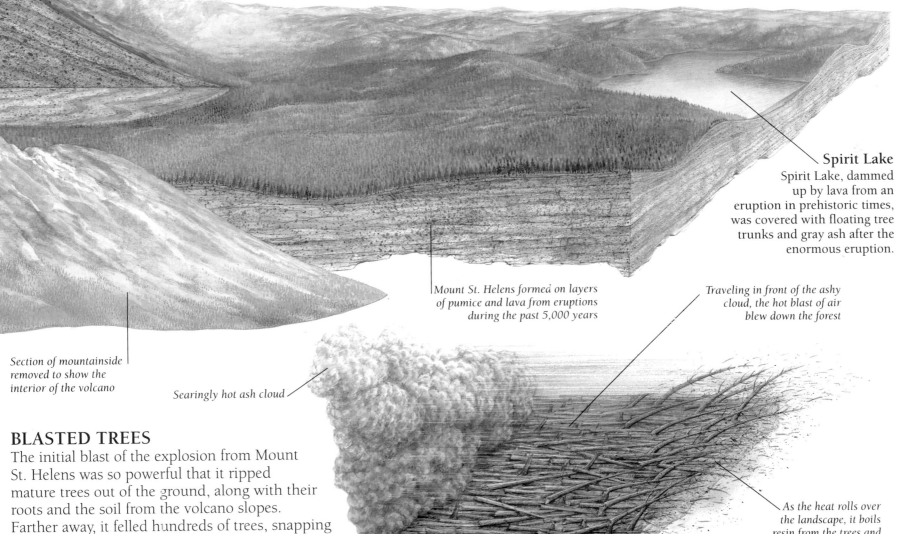

Spirit Lake

Spirit Lake, dammed up by lava from an eruption in prehistoric times, was covered with floating tree trunks and gray ash after the enormous eruption.

Mount St. Helens formed on layers of pumice and lava from eruptions during the past 5,000 years

Traveling in front of the ashy cloud, the hot blast of air blew down the forest

Section of mountainside removed to show the interior of the volcano

Searingly hot ash cloud

BLASTED TREES

The initial blast of the explosion from Mount St. Helens was so powerful that it ripped mature trees out of the ground, along with their roots and the soil from the volcano slopes. Farther away, it felled hundreds of trees, snapping them off just above ground level, stripping their branches, or splintering their trunks as though they were flimsy sticks.

As the heat rolls over the landscape, it boils resin from the trees and turns the water in plants and animals instantly into steam

LAVA ERUPTIONS

IN A VIOLENT ERUPTION, an explosive volcano can devastate the land and throw clouds of searing hot ash into the sky. But other types of volcanoes erupt more quietly and gently, oozing floods of runny, red-hot magma from deep within the Earth's mantle. Because this lava cools to a dark-colored rock called basalt, these volcanoes are known as basalt-lava volcanoes. Basalt-lava volcanoes usually erupt frequently, so they do not build up a huge head of pressure. Instead, the volcano may spit fountains of lava into the air, along with long lava flows that spread out over the surrounding countryside. Over time, these eruptions build huge, broad mountains with very gentle slopes. The islands of Iceland and Hawaii were formed in this way. Volcanic islands grow bigger as lava flows spread into the sea. New islands appear when undersea volcanoes grow large enough to become islands.

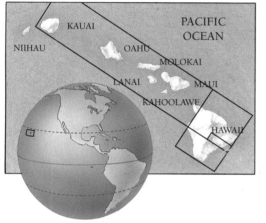

ISLANDS OF FIRE

The island of Hawaii lies in the middle of a great plate underlying much of the Pacific Ocean. Two of the largest and most active volcanoes on Earth – Mauna Loa and Kilauea – are found here. In this illustration, Mauna Loa is split in half to reveal its red-hot interior. Basalt magma from a hot spot in the mantle wells up beneath the volcano, sometimes erupting near its summit. If the magma finds a crack inside the volcano and spreads out sideways, lava may come out from lower on the volcano slopes.

Hawaiian island chain

The Hawaiian Islands are the tops of huge volcanoes rising from the Pacific Ocean floor. Hawaii is the biggest and youngest of a chain of 130 islands. The eight largest islands are seen above.

Sliding sides

Occasionally the side of a volcano falls away in an underwater rock slide. Blocks of island up to 1,650 ft (500 m) across slide deep into the ocean.

HOT SPOT VOLCANOES

The formation of the Hawaiian Islands was a puzzle to geologists, because unlike most volcanoes, they are in the middle of a plate. By determining the age of the volcanoes, geologists reasoned that the entire chain had formed over a stationary "hot spot" deep in the mantle. A volcano forms over the hot spot, but as the plate carrying it moves, the volcano stops erupting and a new, younger volcano forms. In this way, the hot spot builds a chain of volcanoes that increase in age the farther they are found from the hot spot.

The volcanoes grow older the farther they have been carried from the hot spot

Older volcanoes are no longer active

The plate is now moving northwest at about 4 in (10 cm) a year

Path traveled by moving plate

Each volcano stays over the hot spot for about a million years

The youngest volcano forms right over the hot spot, and erupts frequently

Hot spots in the mantle appear to stay fixed in position for tens of millions of years. Some hot spots are found under continents.

This volcano has not erupted for about four million years. The oldest islands are growing smaller as they slowly sink into the sea.

The base of each island is buried in sediment worn from the island itself

Older islands

The islands in the distance are older than Hawaii. They were once above the hot spot, but have been carried away as the Pacific plate moved northwest. Scientists have found evidence for plate movement by taking rock samples from each island. The farther the rocks are found from Hawaii, the older they are.

Hidden depths

Only a small part of the volcano is above sea level. Measured from the ocean floor, the volcano is over 29,530 ft (9,000 m) tall – one of the largest volcanoes on Earth.

Pillows on the seabed

When hot lava erupts under cold water, pillow lavas form. These are rounded blobs of lava with a thin skin. Inside the skin, red hot lava continues to flow until eventually the skin splits and another pillow-shaped blob begins to form. This happens over and over until a jumble of pillows piles up on the seabed.

Fire goddess
Pele, the Hawaiian goddess of fire, is said to have traveled down the chain from the oldest island to the youngest, where she lives today. This myth roughly coincides with the way these "hot spot" volcanoes formed.

BASALT MAGMA
Basalt magma comes from the "guts" of the Earth – called the mantle. As a result, it is hotter and runnier than the type of magma that forms from melting crust at a subduction zone. Only the mantle rocks that melt most easily make up basalt magma; the materials that melt at higher temperatures stay behind in the mantle, making it dense and heavy. When basalt magma comes to the surface, it solidifies to basalt rock. If it crystallizes slowly below the surface, the crystals grow large, forming a coarse rock called gabbro.

At Kilcuea volcano on Hawaii, basalt lava makes a red-hot fountain while lava flows pour over the dark volcanic landscape. Shreds of lava from the fountain collect in a heap as they cool, building a cindery cone.

Crust layers
The lowest layer of the ocean crust is made of coarse-crystallized gabbro rocks. Above this is a wall-like layer formed by magma that has solidified inside dikes. The volcano has grown on top of these layers.

The uppermost mantle is part of the moving plate

The island has been split in two to show the layers of the volcano.

Volcano layers
Because the lowest layers of the volcano erupted under water, they are made up of piles of rounded pillow lavas. Once the volcano broke the surface, the lava erupted on top of the pillows in long gentle flows stretching to the sea.

Each flood of lava covers the older layers formed by earlier eruptions, continuously building the volcano

Under a volcano
Lava that erupts at Hawaii comes from the mantle, well below the bottom of the Pacific Plate.

Lava dikes
The channels that feed magma to the surface are known as dikes. Dikes cut an angled path through the rock layers that they invade.

Loihi, Mauna Loa, and Kilauea are the only active volcanoes in the Hawaiian chain.

Loihi volcano
South of Hawaii lies an active submarine volcano called Loihi. Its summit is 900 m (2,950 ft) below sea level, but one day it will erupt enough lava to become an island.

Loihi is fed by magma from the same hot spot that built Hawaii

Magma curtain
A curtain of magma seeps upward into a crack in a volcano mountain. The magma might eventually reach the surface and seep out as a lava flow. The magma that is not erupted turns into solid rock, forming a dike.

Depth in Miles / Kilometers	
Oceanic crust	0
	3 miles (5 km)
Lithospheric mantle	
	Bottom of plate here
	62 miles (100 km)
Mantle	
	Magma originates here

19

SHAKING THE CRUST

WHEN ROCK MASSES under stress suddenly slip apart and break, the crust trembles in an earthquake. It takes a massive amount of force to snap rock apart. Stress can build up for years in the rock before it is swiftly released. The broken rock moves along a crack called a fault. Most faults are underground, but faults break through to the surface and may even show up in the landscape. The Alpine Fault in New Zealand, shown on these pages, has divided the landscape in two, with rising mountains on the east side, and flat plains to the west of the fault line. Like many faults, it is a result of complex plate movements.

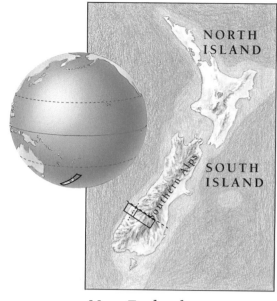

NORTH ISLAND

SOUTH ISLAND

WHY EARTHQUAKES HAPPEN

The three steps below show how stress building in the rocks along a fault line may lead to an earthquake. Two sliding plates lock together, and pressure builds up along the fault line. The stress grows and grows until the strength of the rocks is overcome: the plates suddenly unlock and move, causing an earthquake. The main illustration shows a quake along New Zealand's Alpine Fault.

Stress builds
One plate is moving past another, but the fault line where they meet has gotten stuck. Stress builds up in the rocks of the fault plane, at each side of the fault.

Stress against strength
The stress deforms the rocks, and cracks open in them. Eventually, the stress is greater than the strength of the rocks.

Earthquake
When their strength is overcome, the rocks break, and an earthquake occurs.

Up and down
The New Zealand Alps grow a little higher with each quake, but they are eroded by wind and snow and ice as fast as they are pushed up.

Volcanic landscapes – from geysers to active volcanoes – are found on North Island.

New Zealand
The junction of two plates runs right through the North and South islands of New Zealand, in the Pacific Ocean.

Breaking point
The place where the rock starts to break is called the focus of the earthquake; and the point above, on the Earth's surface, is the epicenter.

The earthquake's vibrations travel out from its focus in all directions.

SOUTH

SOUTHERN ALPS

Alpine Fault

Mt. Tasman

Focus of earthquake

Mt. Cook

Mt. Sefton

Section of main artwork pulled out and made larger

River courses offset (moved apart) along the fault line

Rocks near the surface at the fault are crushed to a greenish rocky mush which erodes easily

Deeper down, rocks in the fault plane may be melted by heat caused by friction

Fault plane
The fault plane is unlikely to be a straight line; it is probably wavy. The irregularities along its length are part of what makes the fault stick between quakes. The more firmly it sticks, the longer time it will be until the next quake, and the bigger that quake will be.

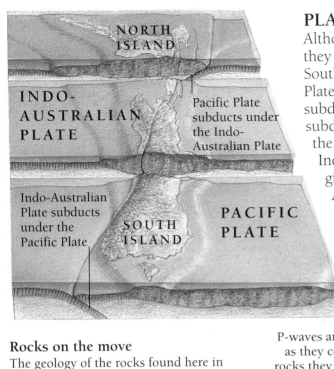

NORTH
ISLAND

INDO-
AUSTRALIAN
PLATE

Pacific Plate
subducts under
the Indo-
Australian Plate

Indo-Australian
Plate subducts
under the
Pacific Plate

SOUTH
ISLAND

PACIFIC
PLATE

PLATE BOUNDARIES

Although earthquakes can strike anywhere, they are most common near plate edges. South of New Zealand, the Indo-Australian Plate is diving beneath the Pacific Plate in a subduction zone. In North Island, subduction is in the other direction, with the Pacific Plate subducting below the Indo-Australian Plate. This results in a gigantic tear through South Island – the Alpine Fault. In North Island, the subduction causes earthquakes, both in the diving Pacific Plate and the overriding Indo-Australian Plate.

San Francisco, California, sits astride the San Andreas Fault. This photo shows streetcar tracks split apart by shaking during a huge earthquake in 1906.

EARTHQUAKE WAVES

The vibrations that travel out from the focus are called seismic waves. These waves move fastest through dense rocks, and more slowly through loose sediments and water. Different kinds of waves cause the rocks to vibrate in different ways. The two kinds that travel fastest are P-waves (primary) and S-waves (secondary). After the vibrations pass, there is no sign of change to hard, solid rocks, but soft sediments may be compacted and pressed together.

Rocks on the move

The geology of the rocks found here in the western plains matches up with that of rocks found on the mountain side of the fault. But because of the sideways movement along the fault, these sets of matching rocks are now hundreds of miles (kilometers) apart. They have been separated by countless earthquakes.

Primary waves

P-waves are like sound waves, as they compress and stretch rocks they pass through. Their simple push-and-pull movement lets the waves travel fast.

Secondary waves

S-waves have a more complicated, shearing motion, moving not only up and down, but sideways in all directions at the same time.

Shaken to the foundations

Shaking ground can collapse buildings, bridges, and other structures. The damage is often most intense nearest to the epicenter.

Sometimes the most damaging vibrations are intensified by soft rocks, where the vibrations travel more slowly.

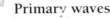

CANTERBURY PLAINS

ISLAND

Lake Tekapo

TSUNAMIS

Earthquakes that occur near ocean coastlines, or that cause the sea floor to slip, can trigger gigantic waves in the ocean water. They are known by their Japanese name, tsunami. Tsunamis have long wavelengths and can travel vast distances, far from the earthquake that causes them – even across the entire Pacific Ocean. When the waves finally meet the continental shelf, the water heaps up into monster waves – on average 100 ft (30 m) high – which can drown a coastline with devastating effects.

Moving north

West of the mountains is a vast plain that has moved at least 300 miles (500 km) north over the last 20 million years. In fact, every time an earthquake hits, the plains move a few inches (centimeters) northward.

The displaced rocks disturb the water above, generating waves

The train of waves travels through the water away from the epicenter

At the continental shelf, the waves break – that is, the water heaps up into huge waves

Waves sweep the shore about every ten minutes

An underwater earthquake causes seabed rocks to break along the fault plane

MOUNTAIN BUILDING

THE EARTH'S SPECTACULAR mountain ranges have been buckled up from relentless plate movements. Most mountains are over a subduction zone at the edge of a continent, but some are the result of uplift caused by huge rifts in a splitting continent. As soon as mountains are lifted up, erosion starts to wear them away. The higher the mountains become, the more rain and snow fall on them – and the more the mountains are carved down by water and ice. Tall, rugged mountains are an indication that mountain building is still taking place. Once there is no more uplift, erosion takes over, and the mountains are worn lower, until eventually only hills remain. Even these low hills contain clues within their rocks to their past as a lofty mountain range.

The Rocky Mountains grew as subducting ocean crust carried islands to the edge of the coast, and packed them together to add to the mountains.

BALANCING FORCES

Most mountain ranges have been shaped by a balance between tectonic (building) forces, which make the region higher, and weathering and erosion, which wear the rocks away. The shapes of the mountains are a result of this balance. Other mountains are built by volcanic activity. A volcanic mountain may sit on top of a tectonic mountain range, making up the very highest peaks. The map on these pages shows the Earth's main mountain ranges.

The folding that formed Sheep Mountain in Wyoming has been revealed by erosion.

South America's Andes mountain range is formed by subduction of Pacific Ocean crust.

Kilimanjaro, Tanzania
An old volcano built of alternating layers of ash and lava, Kilimanjaro is the highest point in Africa at 19,344 ft (5,896 m).

Mt. Elbrus, Russian Federation
Mt. Elbrus is 18,510 ft (5,642 m) high. It is a volcano but it has not been active for many thousands of years.

Mt. McKinley, Alaska
Mt. McKinley stands glacier-clad 20,322 ft (6,194 m) high in the Denali National Park in Alaska.

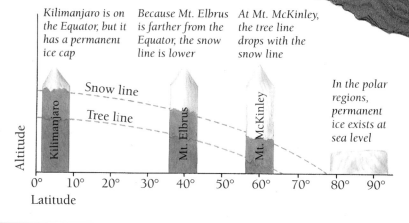

Snow line
The snow line falls lower the farther the mountain is from the Equator.

Kilimanjaro is on the Equator, but it has a permanent ice cap

Because Mt. Elbrus is farther from the Equator, the snow line is lower

At Mt. McKinley, the tree line drops with the snow line

In the polar regions, permanent ice exists at sea level

Tree line
Lower than the snow line is the tree line; above it, trees find it impossible to grow because of the cold.

SNOWY TOPS
Near the Equator, where the direct angle of the sun's rays warms the land, the snow line is over 16,000 ft (5,000 m) up. Only at such heights is it cool enough for permanent ice to exist. Away from the Equator, where the sun's heat rays are more and more slanted, they heat the land less, and the snow line is progressively lower. Sections of three mountains, at different distances from the Equator, are shown here.

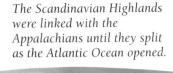

The Scandinavian Highlands were linked with the Appalachians until they split as the Atlantic Ocean opened.

It is so long since the old Ural Mountains were lifted up that erosion has worn them down to only 3,281 ft (1,000 m) high.

The Himalayas are a huge, young mountain range that represents the closing of the old Tethys Ocean.

The mountain ranges in Mongolia were formed 450 to 600 million years ago.

Mountains of the East African Rift Valley have been uplifted as Africa prepares to split apart.

The Alps, Atlas, and Carpathian mountains formed as Africa moved northward against Europe.

The Transantarctic Mountains developed along a rift within the Antarctic continent.

Since the Great Dividing Range formed 300 million years ago, there has been little tectonic activity in Australia.

This satellite image shows the peaks of the Andes Mountains near Santiago, Chile, under their cover of winter snow.

Key to map

Cenozoic mountains, less than 65 million years old

Mesozoic mountains, between 65 and 250 million years old

Younger Paleozoic mountains, between 250 and 450 million years old

Older Paleozoic mountains, between 450 and 565 million years old

MOUNTAINS FROM SUBDUCTION

Some mountain ranges result from ocean crust being subducted under a continent. They are built partly by rising magma from the melted ocean crust, which may erupt or may cool underground as masses of granite.

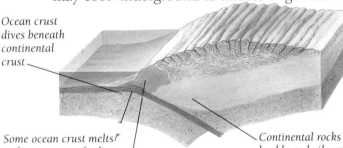

Ocean crust dives beneath continental crust

Some ocean crust melts to form magma, feeding a chain of volcanic mountains that grow with each eruption.

Sediments scraped off the ocean floor are added to the mountains.

Continental rocks buckle and pile up. Some of these rocks are metamorphosed or recrystallized.

Worn-down mountains

America's Appalachian Mountains were built about 250 million years ago, parallel to an even older mountain range. Forces of erosion have attacked the range since its creation, leaving snakelike ridges of hard quartzite rock standing. The softer rocks in between the ridges wear away much more easily, forming valleys. A satellite image of the Appalachians in Pennsylvania (left) shows the distinctive patterns left behind by erosion.

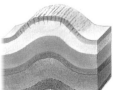

MOUNTAIN STRUCTURES

Making mountains involves both stretching and compression. These illustrations show different types of folding and fracturing that may be seen in mountain ranges.

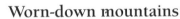

When rocks are stretched apart, they fracture. This type of fault is called a normal fault.

Rocks also fracture when they are compressed. This type of fault is known as a reverse fault.

Rocks crack when they are folded. This allows erosion to wear them away faster. So, the tops of upfolds do not become the tops of mountains.

Folded rocks

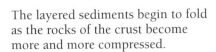

The layered sediments begin to fold as the rocks of the crust become more and more compressed.

Faulted fold

When the folding is so intense that the rocks cannot bend any more, they break, forming a thrust fault.

Folds and faults

In a mountain range, layers of rock are squeezed and folded, then may fracture to form thrust faults. Erosion wears away the tops of the thrust mass of rocks to make the mountain peaks.

CONTINENTS COLLIDE

OVER THE LAST 70 MILLION YEARS, a mighty collision between two continents has created the most spectacular mountain range on Earth – the Himalayas. The continent of India began to move slowly but steadily northward toward Asia, swallowing up an old ocean in its path. Some small continents in the ocean were pushed into Asia first, forming the young Himalayas. As the two continents came closer together, the ocean floor was pushed beneath them, or subducted. When all the ocean had been subducted, India finally met up with the Asian mountains. India has continued to move northward, and Asia has buckled and been pushed up, forming the world's highest mountains and highest plateau.

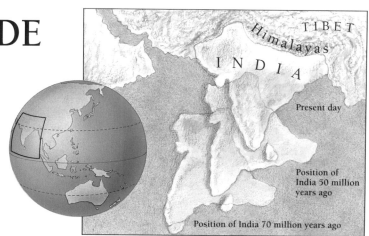

India's northward charge

Before its collision with Asia, the plate carrying India was moving north at 4 in (10 cm) a year. Once the continents met, this slowed to 2 in (5 cm) a year. The Himalayas are the youngest mountains so far formed in this collision.

INSIDE A MOUNTAIN RANGE

India (on the left) has pushed into Asia much like a battering ram. The collision has hardly changed India, but Asia has buckled up and its crust has become almost twice as thick as it was before. This is partly a result of folding and fracturing of the Asian crust, and partly due to blobs of molten rock rising upward to add to the continental mass. As the mountains rise, they are eroded by wind, ice, and rain, making new sediments.

New oceanic crust is formed on the sea floor south of India

Wedge of sediment

This fan-shaped wedge of sediment was scraped off the subducted plate and accreted, or joined, to Asia. It is known as an accretionary wedge.

Young sediments formed from erosion of Himalayan mountains

Old hard continental crust of India

The stark, glacier-hung peaks of Thamserku soar over the Everest region of Nepal.

The roof of the world

The Himalaya range includes the tallest mountain on Earth – Mount Everest – soaring to a height of 29,028 ft (8,848 m). Many other mountains in the range tower to heights of more than 26,240 ft (8,000 m). The exact height of Everest is difficult to determine. This is because the depth of the snow covering it changes all the time.

HIMALAYAS CROSS-SECTION

The rocks that now make up the high mountain peaks of the Himalayas were formed on the floor of the Tethys Ocean, which once separated India and Asia. When the continents met, the heavy oceanic crust was subducted. But the lighter seafloor sediments were scraped off, one slice after another. Through folding and faulting, they became part of the Himalayan mountain range. Faults push one slice of rock over another. This close-up view of a section of the main illustration shows how the rock sequence created by faulting is repeated over and over again.

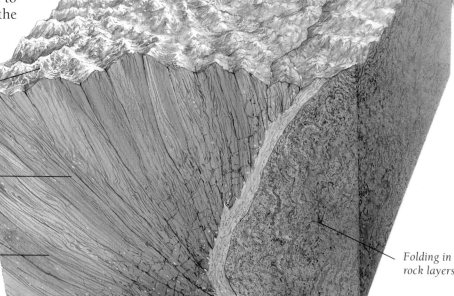

Glaciers keep the mountain slopes steep

Fractures caused by crushing of rock layers

The youngest rocks of one slice are beside the oldest rocks of the next slice

Folding in rock layers

INDIA'S LIFE HISTORY
India became separated from the old southern continent of Gondwanaland when a new ocean formed between it and Antarctica, pushing the landmasses apart. North of India there were several small continents which were added on to Asia as the Tethys Ocean floor was subducted. When India met Asia at last, no more subduction could happen, because continental crust is too light to subduct. Instead the edge of the Asian continent collapsed and was crushed to make room for India.

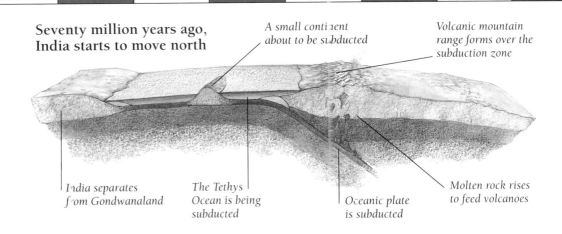

Seventy million years ago, India starts to move north

A small continent about to be subducted

Volcanic mountain range forms over the subduction zone

India separates from Gondwanaland

The Tethys Ocean is being subducted

Oceanic plate is subducted

Molten rock rises to feed volcanoes

Fifty million years ago, India collides with Asia

The last remains of the Tethys Ocean are subducted

The old line (suture) where the small continent added on to Asia

Volcanic activity continues

The southern edge of Asia starts to collapse

Ocean sediments are thrust and stacked on top of each other

Earthquake zone
Deep, violent earthquakes under the west end of the Himalayan mountains show that the subduction process is continuing.

Karakorams
The Karakoram range is home to K2, the second highest peak in the Himalayas at 28,250 ft (8,611 m).

High and dry
Because the massive mountains block the monsoon rains, much of Tibet is a desert.

TIBET

Tibetan Plateau

Karlas Range

Kunlun Shan Mts.

Tien Shan Mts.

TIBETAN PLATEAU
Experts think continental crust may be at its thickest under Tibet. Here, the land surface of the plateau is a lofty 3 miles (5 km) above sea level. The plateau is a stark, treeless land dotted with low mountains. Its plains are thick with sediment worn from nearby mountains. Probably because of fracturing, the mountains seem to be collapsing like wobbly jelly. As a result, the mountains are burying themselves in the sediment they shed.

Tibetan hot springs
There are few active volcanoes over the remains of the subduction zone in Tibet. Tibet does have hot springs, however, which are heated by the cooling igneous rock.

LAKE BAIKAL
The results of the crushing of Asia are visible thousands of miles (kilometers) north, as far away as Lake Baikal, in Siberia. Lake Baikal sits in a huge rift which has been opening for 25 million years. The rift is about 5.5 miles (9 km) deep – almost as deep as the Mariana Trench. Over the years, the rift has filled with sediment, but today Baikal is still the deepest lake on Earth, holding about one-fifth of the planet's fresh water. As India continues to barge northward, Baikal may one day become a new ocean which will split Siberia apart.

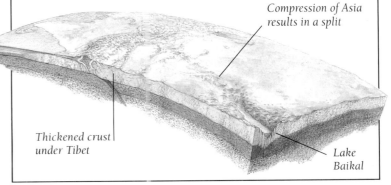

Compression of Asia results in a split

Thickened crust under Tibet

Lake Baikal

Rocks formed by remains of the small continent

Subduction zone
When two plates collide, particularly at the edge of an ocean, the impact can force one plate right underneath the other. This is known as subduction. As the subducted plate dives downward, it starts to melt. Earthquakes and volcanoes are common in subduction zones.

Melted crust
Subducted oceanic crust starts to melt at a depth of about 62 miles (100 km). The fiercely hot molten rock pushes up through any cracks and weak spots in the rocks above. It gathers in huge underground blobs called plutons, which may cool and harden to form igneous rock, such as granite.

MAKING THE CRUST

FORCES BOTH ABOVE and below the ground combine to create, destroy, and change the Earth's crust in a continual cycle. At spreading ridges, basaltic ocean crust is continually being made. This new crust moves away from the volcanic ridge, then plunges back into the Earth's interior at a subduction zone. In the interior, the crust melts into magma. This magma rises toward the surface either to feed a volcano or to form huge underground blobs called plutons. Plutons may cool to make rocks like granite, building the mountain range higher. Deep inside the mountain range, rocks are heated and recrystallized. As the mountains grow, the slopes are worn down and the loose rock fragments go to make new rocks. These processes form the Earth's three rock families: igneous, metamorphic, and sedimentary.

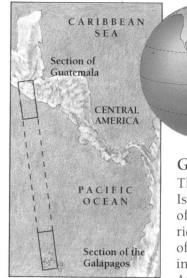

Galápagos
The Galápagos Islands are part of a spreading ridge southwest of Guatemala, in Central America.

CREATION AND CHANGE
The coastal mountains of Guatemala and the neighboring Galápagos Islands are illustrated on these two pages. Some sections have been cut away to show the processes that create the Earth's crust. The mountain range, above a subduction zone, is made partly of granite, an igneous rock made when old oceanic crust melts, and then cools and hardens. As the granite pushes toward the surface during mountain building, some rocks are heated, compressed, and recrystallized to form metamorphic rocks. The rising mountain peaks are eroded almost as quickly as they grow, making sediment that forms new sedimentary rocks near the Guatemalan coastline.

◦The circles show how much of each of the three types of rock – igneous, sedimentary, and metamorphic – is found in each region.

Coastal rocks
Huge amounts of sediment – from large boulders to tiny flakes of clay – wear away from the nearby mountains and settle at the coastline. Only sedimentary rocks are found here.

Mountain rocks
In mountain regions, granites make up about a third of all rock, sedimentary rocks about a third, and metamorphic rocks the rest.

A tor is a craggy hill formed from the erosion of granite. Vixen Tor (above) is in Devon, England.

GRANITE LANDSCAPE
The illustration below shows some features found in a granite landscape. The top of a batholith has been worn away, and nearby are the eroded roots of old volcanoes.

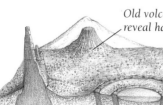

Old volcano erodes to reveal hard lava plug

Tor left after part of batholith erodes

Feeder pipe for a long-extinct, eroded volcano

Collapsed top of a granite pluton fills with water

Part of a huge dome-shaped granite batholith

Sierra Madre Mts.

GUATEMALA

Plutons
The rising blobs of magma are called plutons, after Pluto, the Roman god of the Underworld.

Range rocks
Inside a young mountain range, almost all of the rocks are either igneous or metamorphic.

Granite batholith
A pluton cools slowly over many millions of years to form granite. New blobs of magma may remelt granite that has already cooled, so the history of a batholith can be quite complex.

BASALTIC LANDSCAPE

The illustration on the right shows some of the features of a typical basaltic landscape. Flows of basalt lava spill downhill, over older flows now cracked into columns. On flat land, the lava spreads out in sheets, which may build up in layers. The upper parts of a lava flow are usually covered in blocklike lumps of lava. These weather away relatively quickly, forming a rich soil, which may be reddish in color. Where one lava flow runs over another, a series of "steps" can form in the land. An old word for this kind of landscape is *trap*, as in the Siberian traps.

Columns
These tall basalt columns formed when the inside of an old lava flow slowly cooled and shrank, cracking into matchstick shapes. The rock piles below the cliffs have been eroded from them.

Step in the landscape

Lava flows
A new lava flow cascades over the cliff of basalt columns. The molten lava moves inside a wrinkled skin of cooler lava, at a speed of up to 30 mph (50 km/h). As it cools, the lava slows down.

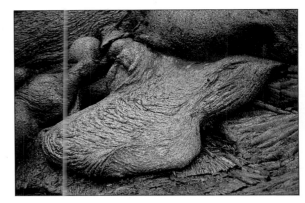
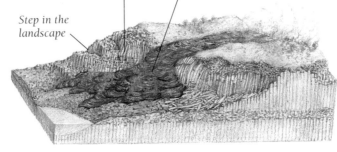

The photograph above shows basalt lava seeping from Kilauea volcano on the island of Hawaii. This hot runny lava with a cooler skin is known by its Hawaiian name, pahoehoe. It is basalt lava that rises at a spreading ridge, building the ocean floor.

Crust destroyed
At a subduction zone, basaltic oceanic crust goes back into the Earth's mantle. The sediments on top of the crust may be scraped off and remain on the edge of the continent.

Ridge rocks
At spreading ridges, almost all of the rock is igneous, with some sedimentary rock.

New seafloor at spreading ridge

Island rocks
On volcanic islands, most of the rock is igneous, but there is some sedimentary rock.

GALÁPAGOS ISLANDS
COLON RIDGE
PACIFIC OCEAN
MIDDLE AMERICA TRENCH

Basalt islands
The Galápagos Islands are built by basaltic rocks, made near the spreading ridge by hot magma rising from the mantle.

Crust created
At spreading ridges, basaltic magma wells up to the surface and crystallizes. Layers of lava spill out on the ocean floor, making new crust, while some magma cools below.

Basaltic rocks go into the mantle and start to remelt at 62 miles (100 km) down

Edge rocks
At the edge of the continent are igneous rocks, and metamorphic rocks that are made from them.

THE ROCK CYCLE
The rock cycle is how rocks change, one into another. A simple diagram of the cycle is shown on the right. Igneous rocks, for example, named for the Latin word for "fiery," form when magma cools underground. They can also form when lava from a volcano cools quickly. Once exposed to the effects of water, wind, and ice, igneous rocks erode, making sediment (from the Latin word for "settle"), which builds up and eventually becomes layered sedimentary rock. In mountain ranges, rock changes due to heat and pressure, and recrystallizes to become metamorphic rock (from the Latin word for "changed"). There is no beginning or end to the cycle, and each transformation may take many millions of years.

Clouds of ash and dust are thrown into the atmosphere

A volcano erupts lava and ash

Igneous rock is weathered and eroded

Ash and grains of rock are dumped in layers

Igneous rock
Extrusive igneous rock forms when erupted lava cools

Magma rises to the Earth's surface and erupts in a volcano as lava

Some sedimentary and metamorphic rock wears away to form new layers

These layers are squeezed and hardened into rock

Plutonic igneous rock forms when magma cools and hardens under the ground

Heat and pressure may recrystallize rock into other rock

Sedimentary rock

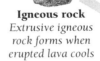
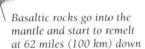

Metamorphic rock

Rock melts to form magma

Rock that is heated enough may melt to make new magma

Igneous rock

THE CRUST WEARS DOWN

AS NEW CRUST is created, forces act upon it all the time to wear it down again. These forces might be as obvious as a gigantic landslide, or as invisible as a flake of clay carried in a raindrop. Changes in the weather bring continual fluctuations in temperature and dampness. These changes affect surface rocks. The rocks expand and contract, and become waterlogged and dried out. Mineral grains that make up the rocks loosen and separate, creating many tiny rock fragments. The fragments may remain in place or be carried away by rain, by melting snow and ice, by the wind, or by rivers, such as China's mighty Huang He, seen here. Plant roots play their part by wedging rocks apart along cracks, allowing water to penetrate more deeply. Weathering is the breaking apart or chemical "rotting" of the rocks, and erosion is the removal and transport of rock grains.

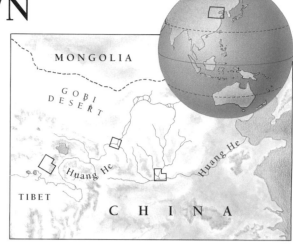

Course of the Huang He
The Huang He (Yellow River) of China travels 3,000 miles (4,830 km) from its source in the mountains of northern Tibet, through a loop near Mongolia, and south to the Yellow Sea.

The Huang He is so loaded with yellow silt that it is the same color as the silt of the riverbanks.

Meandering to the sea
Nearer to the sea, the river travels over the gentle slopes of the flood plain. By this point the river is loaded with yellow sediments. As it meanders across the land here in great, slow-moving curves, it drops much of its load of silt. The river flows between banks made of silt that has already been deposited. Each year, the river drops more silt, filling up the river channel.

Turning yellow
The river water is clear in the rocky mountains. But the soft, fertile yellow silt of the loess region is easily washed away, especially from plowed fields. Here the river picks up the bulk of the yellow sediments that give it its color.

When it leaves the loess lands, almost a third of the river's total volume is sediment

Rich flood plains
For many thousands of years, flooding has been disastrous to the people living here. Yet it is the sediments deposited by the flood waters that enrich the soil and draw people to the area.

Slower-flowing river drops its load of sediment

Old layers of silt and gravel are laid beside one another on the inside of the bend

A CURVING COURSE
As water rounds a curve in the river course, it flows fastest on the outside of the bend. The river is also deepest here. On the inside, the water is shallower and flows more slowly, so silt and pebbles are dropped. Over time, the river bed moves outward, making a gentle bend into a sharp curve. In this way the course of the river moves sideways as it snakes across flat land.

Fast-flowing water on the outside of the bend

Deeper water erodes the sand and gravel banks

CHINA'S HUANG HE

Three sections of the Huang He are shown on these pages. The river carries some sediment from its mountainous source in Tibet. Farther downstream is the loess region, where soft silts are easily weathered and eroded. The river picks up the fine, silty loess grains and carries them away. In the flat plains near the sea, the river flows slowly and can no longer carry the silt, which drops to the riverbed, building it ever higher.

Crumpled peaks
The Kunlun Shan (or mountains) are eroded into rugged peaks. As the rivers cut valleys ever deeper, the peaks become steeper and more unstable.

Ups and downs
The higher the mountains are pushed up by forces within the Earth, the faster erosion wears them down.

The mountain region
In winter, snow and glaciers cover the high mountains. The mountain rocks may be broken apart as water in cracks within them freezes at night and thaws during the day. In summer, the snow melts to feed fast-flowing torrents. These torrents carry the loose rock fragments swiftly downstream.

In the mountainous regions, turbulent streams flow over rocky valley floors

Valleys have steep sides

Carving a valley
These river valleys have steep sides because layers of very hard rock make up the mountains. Where the underlying rock is softer, valleys are less rugged.

HILLSLOPE EROSION

The pull of gravity means the surface on almost every slope is slowly inching downhill. Slope erosion happens most rapidly on steep hillsides that are unstable or barely stable. Gentle slopes and tree-covered slopes are less easily eroded.

Falling rock
At the foot of steep, rocky slopes there is usually a heap of rock fragments that have fallen from the cliff face. This heap, or talus, makes the slope less steep and reduces the rate of erosion.

Enlarged cross-section of area in the loess

Layers of pebbly silt are soft and easily eroded

Slipping land
Steplike landslips happen along curved cracks in the rock. These too make the slope more gentle, unless the "toe" at the bottom of the slip is eroded by a river in the valley.

The loess lands
The hill slopes in the middle course of the Huang He are rapidly eroding. These slopes are made of soft loess, young layers of wind-blown silt that were blown from the deserts of central Asia and deposited here during the Pleistocene Ice Age. The river valley is flat in places where the valley is wide and layers of sediment line the riverbed. In others parts, where the river goes through harder rocks, its silty water flows in torrents through steep-walled gorges.

A series of steps called terraces have been cut into the slopes. These allow farmers to grow crops on level ground right up to the hilltops, and help prevent heavy rains from washing the soil away.

Tree grows up toward sunlight despite the sliding soil

Sliding soil
Even though its movement may be too slow to see, all slopes show some evidence of the continual downhill creep of soil. Each rainstorm and frost moves soil and broken rock fragments in the subsoil slightly farther downhill. Heavy rainfall can wash whole slopes clear of soil and talus, especially where there are few trees to hold the loose slope material in place.

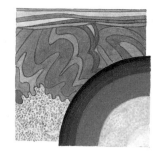

LAYERING THE LAND

THE SEDIMENT that is carried by rivers and glaciers from mountain regions eventually finds a resting place. This might be as boulders and gravel at the foot of a mountain, sand dunes in a desert, silt and salt in a desert lake, or as sand or pebbles on a riverbank or on a coastline. Sediments are laid down in layers, each younger layer on top of older ones. These layers are exposed to view where they have been uplifted and then cut through by eroding rivers. Perhaps the most spectacular example on Earth is the immensely deep Grand Canyon, carved into the layered rocks of the Colorado Plateau by the Colorado River.

This spectacular view of the Grand Canyon was taken from Mather Point on the South Rim. The higher North Rim is just visible in this picture.

North Rim
More than 1,000 ft (300 m) higher than the South Rim, the North Rim is covered with snow until late spring.

Weathered walls
Over millions of years, weathering has chiseled and shaped the curves and gulleys of the canyon walls.

Canyon colors
The layers of different types of rock give the canyon its spectacular colors, including gray limestone, yellow sandstone, pink granite, and black schist.

Bright Angel Point

BRIGHT ANGEL CANYON

KAIBAB PLATEAU

Widforss Point

Manu Temple

Buddha Temple

HAUNTED CANYON

Cheops Pyramid

Tiyo Point

Isis Temple

HOW COAL FORMS
Coal began to form when vast swampy forests flourished on a river delta. As the trees died, they fell into the swampy bogs. Instead of decaying as they would in air, the dead tree remains were preserved under the water, sticking together to form a dark, fibrous material called peat. This was buried by layers of new sediment: sand, lime, and mud. As more and more sediments were laid down, each peaty layer of forest remains became compressed. After many millions of years, the peat was compressed and heated, forming coal.

Coal deposits, or seams, are removed from the Earth by mining.

Surface works of coal mine

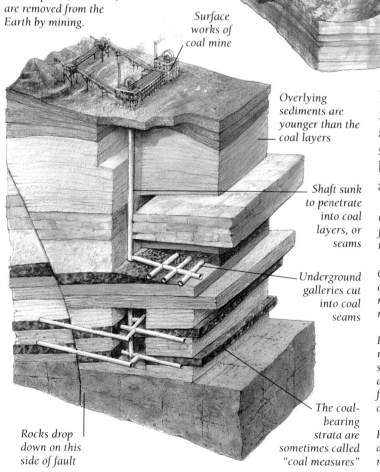

Overlying sediments are younger than the coal layers

HOW OIL FORMS
Some sediments contain large amounts of the remains of tiny sea plants. When these are buried, they are "cooked" by heat and pressure to become oil.

Shaft sunk to penetrate into coal layers, or seams

Oil is pumped from the rocks to an oil rig

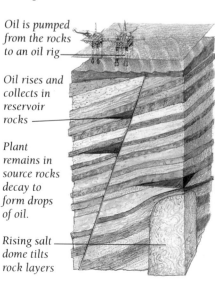

Oil rises and collects in reservoir rocks

Underground galleries cut into coal seams

Plant remains in source rocks decay to form drops of oil.

Trees to peat
Soft, nonwoody trees that grow in a tropical swamp may form peat.

Peat to coal
Layer after layer of new sediment compresses the peat to coal.

Rocks drop down on this side of fault

The coal-bearing strata are sometimes called "coal measures"

Rising salt dome tilts rock layers

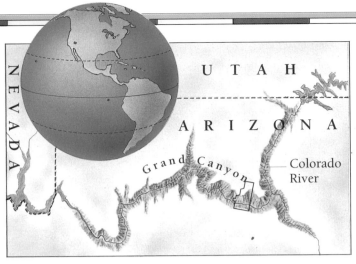

CHANGING SEDIMENTS

As sea level changes with time, so coastlines move. Sea level rises when spreading ridges are more active. The undersea mountains push aside ocean water, causing it to flood the continental shelves.

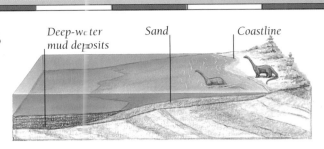

CHANGING SHORELINES

At times when sea level is lower, during an ice age, or when spreading ridges are less active, the continental shelves dry out. Shoreline sediments such as sand are then deposited on top of the older, deeper water sediments.

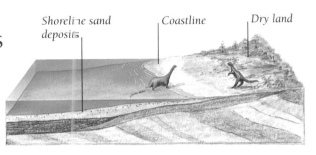

Arizona's Grand Canyon and Colorado River

The Colorado River has carved the awesome Grand Canyon through Arizona, in the southwestern United States. The canyon is 217 miles (349 km) long, up to 18 miles (30 km) wide, and 1 mile (1.6 km) deep.

Rocks of the rims

The rocks near the canyon rims were formed from sediments laid down 250 million years ago. Each layer reflects the changing history of the area. Some layers were once ancient seabeds, while others were desert sands.

South Rim

Little rain falls on the desert plateau on the South Rim, so there is little water to run down and erode the canyon walls. As a result, the South Rim of the canyon has the steepest slopes, and the river runs right at the foot of the cliff walls.

Fossil finds

Fossils found in the canyon walls record a changing history of life. Older life forms such as trilobites are found near the bottom of the canyon, while reptile fossils are nearer the top.

Rockfalls widen the upper part of the canyon.

Yaki Point

O'Neill Butte

GRANITE GORGE

Mather Point

Bright Angel Creek

Yavapai Point

COCONINO PLATEAU

The Battleship

TONTO PLATFORM

Powell Point

Hopi Point

Colorado River

Hopi Wall

TONTO PLATFORM

Great Mohave Wall

The rock fragments carried by the river wear away new sediments as the river runs through the canyon

HISTORY OF THE CANYON

At its deepest point, the Grand Canyon cuts through rocks 2 billion years old. The oldest, metamorphic rocks at the bottom of the gorge were formed as part of an ancient mountain range. Younger layers of limestone, shale, and sandstone were laid over these mountains, then lifted up into the Colorado Plateau. As the plateau rose and domed up in the middle, the river carved its course deeper, keeping pace with the uplift. The rocks of the canyon rim are about 250 million years old.

The Colorado River

From its source in the Rocky Mountains of Colorado, the river falls sharply through several canyons and into the desert. The Colorado River itself may be as much as 30 million years old, while the high plateau itself is younger.

Deep cut

The rising plateau gave the river a steeper slope on its journey to the sea. This made the river flow faster, so it had more energy to cut away its riverbed and deepen the canyon floor.

Steep cut

The plateau's rapid uplift means the river has carved an even steeper canyon over the last two million years. In its long history, the canyon may never have been so impressive as it is today.

THE CRUST CHANGES

AT THE HEART of the Earth's mountain ranges lie metamorphic rocks. They are different from sedimentary and igneous rocks – they have unique textures and structures, and contain new and different minerals. Metamorphic rocks are igneous or sedimentary rocks that have been changed. The changes are brought about by heat given off at nearby igneous rock intrusions, by the immense pressure from the weight of the overlying mountains, or by chemical reactions. Heat and pressure cause the rocks to recrystallize without melting. Metamorphic rocks in a landscape show there was once a mountain range there, though there may be no high ground any longer.

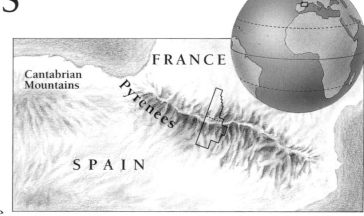

The Pyrenees Mountains
Forming a border between France and Spain 270 miles (435 km) long, the Pyrenees are one of the Earth's young, still growing mountain ranges.

METAMORPHIC MOUNTAINS
The Pyrenees mountain range, seen on these pages, has granite intrusions at its core. The hot molten granite swelled upward, heating the surrounding rocks and pushing them aside. Over many millions of years, the sedimentary rocks and older igneous rocks were metamorphosed within the mountain range. Today, many miles (kilometers) of rock have been eroded off the top of the Pyrenees, revealing its metamorphic heart.

These layers of sedimentary rock at Pico de Vallibierna have been folded and over-folded until they are lying down, or recumbent. Part of the fold is eroded.

Pico de Vallibierna, Spain
The folded rock pattern shown in the photograph on the left is outlined here.

Folding up rocks
In mountain ranges, sedimentary rock which was first laid down in horizontal layers is folded. The forces that make the mountain range push and fold the layers, sometimes forming intricate patterns later exposed by erosion.

Change in the range
Metamorphism that happens over a wide area, especially in mountain ranges, is known as regional metamorphism.

Lago Helada, Spain
The outline shows a recumbent fold in the strata of Lago Helada, seen on the right. The folding is so intense that some of the layers are now upside down. The top of the fold has been cut away by erosion.

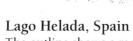

This icy lake in Spain's Ordesa National Park shows a reflection of the folded rocks of Lago Helada above.

Section through granite intrusion

Intrusion
This granite intrusion cooled many miles (kilometers) down in the crust. It is now exposed at the Earth's surface by erosion.

Country rock round intrusion

HISTORY OF THE PYRENEES

A hundred million years ago, the great Tethys Ocean stretched between Europe and Africa. As immense forces in the Earth pushed Africa northward, the ocean was swallowed at a subduction zone. Sediments from the ocean floor were buckled and crushed and intensely over-folded, as great rock masses were pushed one over another to make the Pyrenees. Some rocks deep in the mountain range were heated, just by their deep burial. These rocks became hot enough and were buried for long enough to recrystallize into metamorphic rocks.

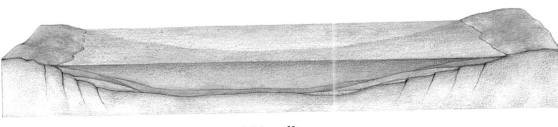

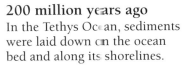

200 million years ago
In the Tethys Ocean, sediments were laid down on the ocean bed and along its shorelines.

100 million years ago
Iberia and France crunched together, and a mountain range began to grow.

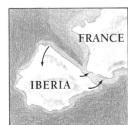

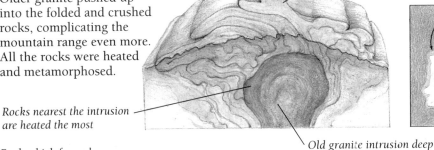

Rocks are folded and pushed one over another

Two million years ago
Older granite pushed up into the folded and crushed rocks, complicating the mountain range even more. All the rocks were heated and metamorphosed.

Rocks nearest the intrusion are heated the most

Old granite intrusion deep within the mountains

IBERIAN DRIFT

As Africa moved northward, Iberia (the landmass of Spain and Portugal) crunched sideways against Europe. Layers of sediment from the old ocean floor were probably scraped up and eventually eroded away, so there is no record left of their existence. Today, the Mediterranean Sea is all that remains of the ancient Tethys Ocean.

Ice carving
The high slopes of the Pyrenees have been carved into rugged peaks by the glaciers that covered them during the Pleistocene Ice Age.

Shared history
The Pyrenees are among the world's youngest mountain chains, the same age as the Alps and Carpathians in Europe and the Himalayas in Asia.

Rock which formed many miles (kilometers) deep has been worn from the peaks

Folded and metamorphosed rocks near the granite

Foothills
The rocks of the foothills show less complicated folding and less intense metamorphism.

The dark rock at the center of Pic la Canau has eroded more easily, making a gully on the slope.

Pic la Canau, France
The outline shows how the rock was stretched and cracked over the top of the fold, allowing erosion to bite in. The top of the fold has all worn away.

HOW ROCKS CHANGE

As rocks are compressed and heated, their mineral grains gradually rearrange themselves in response to the changing pressure and temperature. Atoms move from places where pressure is greatest to the places where it is less intense. New minerals grow which are stable at high pressures and temperatures. The type of metamorphic rock that forms depends on the composition of the original rock, and how much heat or pressure – or both – brings about the change.

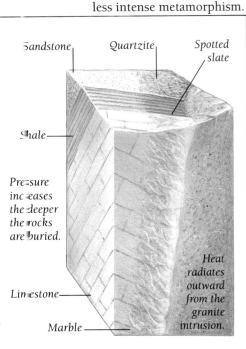

Sandstone
Quartzite
Spotted slate
Shale
Pressure increases the deeper the rocks are buried.
Limestone
Marble
Heat radiates outward from the granite intrusion.

33

AMAZING EARTH

EARTH'S LANDSCAPES show an amazing variety – from deserts and polar ice regions to spectacular waterfalls and graceful volcano cones. Each has its own unique geological history. Some landscapes are the result of recent erosion or tectonic (building) processes, which in the vast time scale of Earth's history means changes over the past few tens of millions of years. Others were sculpted by the same processes, but have changed little for hundreds of millions of years. Several amazing landscapes – from hoodoos to inselbergs – are shown here.

Bryce Canyon, Utah

The hoodoos (from an African word meaning "spirit") of Bryce Canyon are a mass of pinnacles sculpted from layers of soft young rock. The canyon's pink-orange limestone is sediment that collected in a lake only 60 million years ago. The attack of wind, snow, and rain has worn the rocks into colorful hoodoos.

These red sandstone towers are the North and South Mittens.

Loose rock collects at the base of the hoodoos.

Monument Valley, Utah

The large mesas and smaller buttes that tower over Monument Valley are isolated flat-topped mountains, made of horizontal layers of sedimentary rock. Over hundreds of thousands of years they have worn away, leaving behind tall towers of rock.

The falls plunge off a flat plateau.

The Pantanal, Brazil

In the back country of Brazil, seasonal rainfall in the mountains feeds mighty rivers. Where these rivers travel over the level swamplands of the Pantanal, they spread out, flooding the land. When the rains stop, hundreds of shallow pools are left behind.

The swamps cover an area the size of Great Britain.

Angel Falls, Venezuela

The waterfall with the longest drop in the world tumbles 3,212 ft (979 m) off the wet swamplands of a plateau called Auyán Tepuí in Venezuela. It is named after the pilot Jimmy Angel, the first outsider to see the falls in 1935. The water changes into white mist before reaching the bottom.

Canadian tundra

In summer, soggy plains stretch in all directions in the Arctic regions of northern Canada and Siberia. Below the surface the ground is permanently frozen, so the summer meltwater has nowhere to go and collects in swampy pools. At the end of the summer these pools of water freeze again. When water just beneath the surface expands to form ice, it may push the soil up into small domes called pingoes.

Tundra
NORTH AMERICA
Bryce Canyon • Monument Valley
Tropic of Cancer
ATLANTIC
Equator
Angel Falls •
PACIFIC OCEAN
SOUTH AMERICA
Tropic of Capricorn
The Pantanal •

Antarctic ice cap

A vast sheet of ice makes up the cold deserts of Antarctica. The ice cap has formed from frozen snow that has accumulated over tens of thousands of years. Covering nearly the entire continent, the ice is over 14,764 ft (4,500 m) thick in places. Only the tallest summits of the Transantarctic Mountains break through the ice.

The icecap contains 90 percent of all the ice on Earth.

Ahaggar Mountains, Algeria

From the desolate Sahara desert plain rise the majestic Ahaggar Mountains. The tallest of these spiny peaks is about 9,840 ft (3,000 m) high. The mountains are made of igneous rocks – granites and lavas including phonolite. Phonolite, meaning "sound stone," is so called because when it is hit with a hammer, it gives off a musical note.

Fuji is a sacred place of pilgrimage. Thousands of people each year climb the mountain to watch the sun rise.

The phonolite cooled and cracked into long, thin shapes that give the Ahaggars their ribbed surface.

Mount Fuji, Japan

The majestic snow-capped volcano Mount Fuji is 12,388 ft (3,776 m) high. The volcano has been active for thousands of years. When it last erupted in 1707, black ash fell in the streets of Tokyo, 62 miles (100 km) away. Its name comes from "fuchi," which means fire, a word of the Ainu, the original people of the Japanese islands.

Guilin hills, China

Over hundreds of millions of years, the limestone in the hills of Guilin has been slowly dissolved by rain, creating a landscape called tower karst. The flat lands at the bottoms of the hills, covered by rice paddy fields, are layered with vast amounts of clay washed away with the limestone. Rivers snake their way around these strange, weathered remains.

Table Mountain, South Africa

The sandstone layers of South Africa's Table Mountain were laid down 500 million years ago. Over time, the sand hardened into rock and was uplifted without folding, so its layers are still horizontal. Erosion has worn away everything but the distinctive table rock that remains.

Kata Tjuta, the Aboriginal name for the Olgas, means "many heads."

The Olgas, Australia

Resembling huge red rock haystacks, the Olgas (or Kata Tjuta) are clustered on the sandy Australian plains. The plain is covered with regolith – rock-sand and clay weathered from the underlying solid rocks. Erosion does not remove the regolith, so over time it gets thicker, until it is burying all but the highest points of the underlying solid rock. These island mountains are called inselbergs. Uluru (formerly Ayers Rock) is another example.

Table Mountain rises 3,566 ft (1,087 m) above Cape Town.

PLANET WATER

EARTH IS ENVELOPED in a watery shroud, with almost three-quarters of its surface covered by ocean waters. The largest ocean, the Pacific, is also the oldest, and is about one third of the Earth's entire surface. Around its fringes lie the deepest places on the Earth's surface, the great ocean trenches. Some trenches are much deeper even than the nearby land mountains are high. The ocean floor is studded with volcanoes and flat-topped mountains that may rise much higher than those on land, and is crisscrossed by massive mountain chains. Layers of sediment carpet the entire deep ocean floor. Shallow water surrounds each continent; these continental shelves are the flooded margins of the continents.

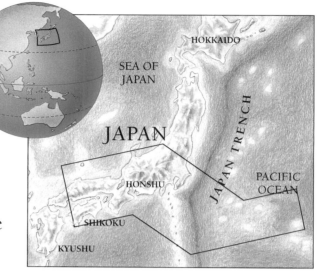

Islands of the rising sun
Japan is a chain of islands east of Asia. Its name means "land of the rising sun" in Japanese. A deep ocean trench lies along its Pacific coast, and a small young ocean separates it from Asia.

At the edge of the continent
The continental slope stretches from the edge of the continental shelf all the way down to the ocean deep. It is covered with layers of mud, sand, and the fine sediments eroded away from the nearby continent.

Seabed sediment
Thick layers of sediment collect on the shallow seabed. Some is made from the shells of sea creatures, and some is sand and mud from the continent.

Young mountains
High mountains with steep slopes make up much of Japan's landscape. The young mountain rocks erode rapidly, pouring loose sediment into the ocean.

Loose sediment settles in layers on the edge of the slope

JAPAN TRENCH
Japan lies on an active continental margin, where the Pacific Ocean plate is being subducted underneath Japan. The deep Japan Trench shown on these pages marks the place where the ocean floor plunges beneath its neighboring plate. As the plate crunches underneath, earthquakes are generated. Sometimes the shaking dislodges loose sediment from the sides of the trench, which then cascades down into deeper water.

Sea canyons
Undersea avalanches of muddy sediment carve out steep-sided canyons into the depths of the trench.

PASSIVE MARGINS
The continental shelves fringing most land areas in the Atlantic Ocean are wide and shallow. They are known as passive margins. As Europe and the Americas split and the ocean between them grew, the continental edge was pushed from the spreading ridge. Much of the continental shelf was flooded at the end of the Ice Age.

An old avalanche of sediment from the coastal water

Ocean trench
A trench is deep and narrow, with steep walls. Some of the sediment that collects here is carried back into the Earth with the subducting plate.

Passive margins have no subduction zone, and few earthquakes

Continental slope ends at a pile of sediment called the continental rise

Continental shelf is part of the neighboring continent

LAND HEMISPHERE

Although Africa, Asia, and Europe dominate this view of the globe, they are surrounded by the waters of the Atlantic, Indian, and Antarctic oceans. The Atlantic is a young ocean, split down the middle by a huge underwater mountain chain. The Atlantic and the Antarctic oceans have been growing larger over the past 200 million years, as the continents have drifted apart. At the same time, an old ocean that once separated the northern and southern continents has shrunk to become the Mediterranean Sea.

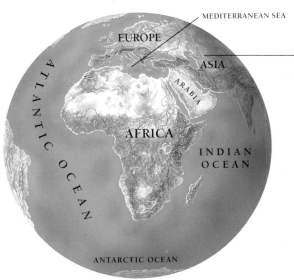

The Aral Sea is a landlocked salty lake, which straddles Uzbekistan and Kazakhstan in western Asia. Diversion of its water for irrigation has made the lake shallower, so the water evaporates faster. These ships are stranded in the shrinking sea.

OCEAN HEMISPHERE

The Pacific Ocean is so vast it extends across a whole hemisphere, stretching nearly halfway around the globe at its widest point. Within the Pacific is the lowest point on Earth – the Mariana Trench. The Pacific is dotted with more than 25,000 islands, which are actually huge mountains rising from the ocean floor. Although subduction has been swallowing the ocean crust along its margins for many tens of millions of years, the spreading ridges to the south and east of the ocean, where new crust is formed, are keeping pace.

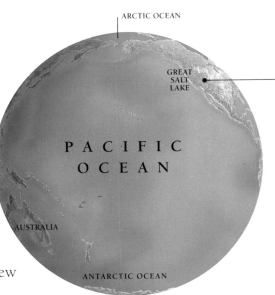

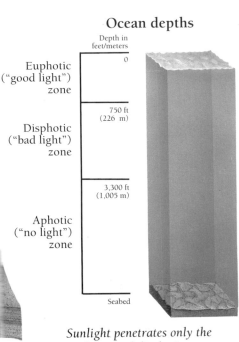

This is a satellite view of the Great Salt Lake and Desert in Utah. No plants can grow in the vast salt flats north of the lake. The lake looks as if it is two colors because it is split by a rocky causeway. A river feeds the left side of the lake, but the fresh water cannot circulate to the saltier right side.

Undersea volcanoes
Volcanoes pepper the ocean floor. Some are active and grow large enough to rise above the water, forming islands.

Guyots
A guyot is a sunken flat-topped mountain. Guyots were once islands, but their tops were worn flat by pounding waves. They sank under the sea as the sea floor subsided.

Age of the seafloor
The ocean floor being subducted under Japan today is about 200 million years old. It was once part of Panthalassa, a massive ocean nearly twice the size of the Pacific that surrounded Pangaea.

Ocean depths

Depth in feet/meters

Euphotic ("good light") zone — 0

750 ft (226 m)

Disphotic ("bad light") zone

3,300 ft (1,005 m)

Aphotic ("no light") zone

Seabed

Sunlight penetrates only the euphotic and disphotic zones. The oxygen-rich water supports a vast range of life. The aphotic zone is inky black with cold waters and immense pressure –few life forms can survive here.

Tipping toward the trench
Flat-topped guyots tip over at an angle as they are pulled into the ocean trench by the subducting ocean floor.

Sediment
Layers of sediment made mostly of red-brown clay build up on the ocean floor.

Final resting place
Buried in the deep sea sediment are whale earbones, skeletons of microscopic creatures - even bits of meteorites from space.

THE OCEAN FLOOR

THE DEEP OCEAN FLOOR is crisscrossed by the longest mountain chains on Earth. These are the spreading ridges, where magma oozes up to form new oceanic crust. The ocean floor today is the youngest part of the Earth's crust. It has all formed at oceanic spreading ridges over the past 200 million years. No older ocean crust is left, because it has all been swallowed back into the mantle at subduction zones near the continental margins. New crust is made at cracks in the spreading ridges. When the crust beneath the cracks stretches, rift valleys form. As the continents on either side of the rift valley move farther apart, there is continually space for more and more new crust. Tall chimney stacks called black smokers billow thick black clouds along cooler parts of the spreading ridge.

RIDGES AND RIFT VALLEYS

A section of the vast underwater mountain chain that snakes through the Atlantic Ocean, the Mid-Atlantic Ridge, is shown here. Its massive peaks rise up to 13,000 ft (4,000 m) above the ocean floor. The rift valley at the center of the ridge is where the ocean floor is spreading. Several black smokers dot the middle of the valley. A section of the rift valley is pulled out and shown larger, to reveal its structure.

Faults

As the growing ocean floor stretches, it cracks along lines (faults) more or less parallel to the rift valley. Steep cliffs along the fault lines gradually grow less steep as blocks of rock break off and fall to the bottom.

Each section of ocean floor breaks along a sloping fault, so its layers become tilted

A black smoker grows when the boiling water shooting from a crack meets the cold water near the ocean floor

The rising cloud of water looks black because of the minerals it contains

The chimneys are brittle and break off from time to time, leaving heaps of broken fragments around the smokers

Pillow lava

Magma oozing up from the mantle becomes basalt lava, a dark-colored rock rich in iron and magnesium. When the hot lava cools in contact with seawater, it makes lumpy round shapes called pillow lava.

Dikes

Under the pillows is a layer of vertical dikes, where magma crystallized as it came up through the rift valley crack.

BLACK SMOKERS

These tall stacks are made of mineral deposits. The black "smoke" they belch out is actually tiny grains of metal sulfides. Originally, the sulfides were in the new rocks of the ocean floor. As ocean water seeps through cracks in the cooling rock, it dissolves the sulfides. Hot magma below the center of the rift makes this water boil. As the water bubbles up through the cracks, it deposits the sulfides and other minerals, building chimneys up to 33 ft (10 m) tall.

Mysteries of the deep

A hundred years ago, the unexplored deep waters of the ocean were a great mystery. The instruments and ways to study the seabed had not been invented. Instead, many people believed folktales of fantastic undersea cities, or fearsome sea monsters prowling the icy depths. In those days, the deep ocean was the setting for science fiction books, such as Jules Verne's *Twenty Thousand Leagues Under the Sea* (left), in the same way as space and other planets are the setting for fantasy novels today.

MID-ATLANTIC RIDGE

Two hundred million years ago, there was no ocean between Europe and Africa, and the Americas. Then a crack developed which grew and widened, and new ocean crust filled in the gaps along the Mid-Atlantic Ridge to form the Atlantic Ocean. By dating ocean-bed rocks, scientists know that the oldest oceanic crust is nearest to the continents, and "strips" of ocean floor are younger the nearer they are to the still-active ridge. The map below shows the age of the crust in each of its sections.

The longest mountains on Earth

The Mid-Atlantic Ridge stretches 7,000 miles (11,500 km) from Iceland in the north to the edge of the Antarctic Ocean in the south.

Transform faults

The spreading ridges are in short sections across the oceans. Every few tens of miles (kilometers), the active part of the ridge is moved sideways by fractures called transform faults. These faults cut across the middle of the ridge and offset the crust into sections.

The fractures extend far beyond the transform fault, where they offset the spreading ridge

Piles of pillows

Although the outside of a pillow lava cools rapidly to a dark skin, red-hot magma still flows inside. It may break through the skin to form a new pillow, so that great heaps of pillow lavas pile up to make the uppermost part of the ocean crust.

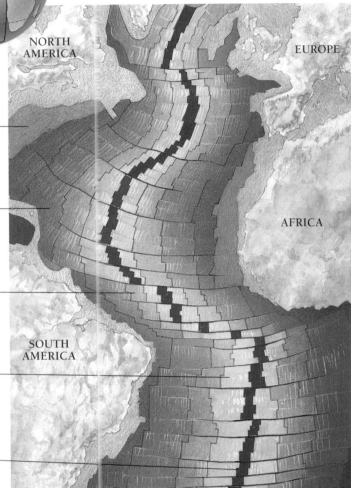

200 million years ago (MYA)
The oldest crust is nearest the land.

65–140 MYA
The crust here formed in the age of dinosaurs.

20–65 MYA
The Pyrenees and Himalayas grew as this crust formed.

2–20 MYA
The Himalayas soared higher as this crust formed.

0–2 MYA
The youngest crust is nearest the ridge.

NORTH AMERICA
EUROPE
AFRICA
SOUTH AMERICA

Lifting the crust

Cracks develop as magma pushes up against the overlying continental crust, stretching and lifting it upward.

Rift valley forms

The continent begins to crack apart, and a central block sinks, forming a rift valley. Magma squeezes in to fill the cracks.

From valley to sea

Fresh cracks fringe the widening rift valley as its walls move farther apart. Seawater enters the deep basin.

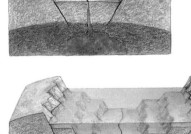

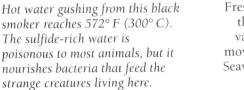

Hot water gushing from this black smoker reaches 572° F (300° C). The sulfide-rich water is poisonous to most animals, but it nourishes bacteria that feed the strange creatures living here.

BIRTH OF THE RED SEA

The Red Sea in eastern Africa formed in the same way as the Atlantic Ocean – at a spreading ridge. Today it is only about 186 miles (300 km) across, but millions of years from now it may rival the Atlantic in size. The Red Sea was born 20 million years ago as Arabia started to crack away from Africa. First, great layers of basalt lava poured out over the surrounding land through cracks in the crust, as magma rose from beneath. Later, the basalt eruptions were concentrated in the central part of the widening rift. This basalt eventually became the new ocean floor as seawater filled the valley. The Red Sea will keep getting wider as long as Arabia keeps moving away from Africa.

THE LIFE OF A RIVER

A TINY TORRENT OF WATER high up in the mountains eventually becomes a vast, placid river flowing into the sea, but it has to go through several stages and pass through many landscapes before it gets there. Streams and rivers are fed from water that runs over the surface of the ground as well as from underground water seeping out from the rock. In the great continents, rivers usually rise in mountains near one coastline. Some travel vast distances across ancient routes to a far coastline; others take their water to inland lakes or seas. The river shown on these pages is the Nile – the longest river in the world.

Course of the Nile

The Nile River flows 4,145 miles (6,670 km) northward through the Sahara Desert to the Mediterranean Sea. The shorter route eastward to the Red Sea is blocked by a range of mountains that are younger than the river course.

BIRTH OF A RIVER

In mountain regions where there is plentiful rain and melting snow, rivers are turbulent (rough). Many little rivulets flow over steep rocky landscapes, carving their pathway as they go, and cutting valleys ever steeper and deeper. Some mountain ranges today are slowly growing higher, pushed up by forces inside the Earth. River valleys in these growing mountains stay steep and full of waterfalls. The rising mountains ensure plenty of rain and snow from cooling weather clouds, which keep the rivers flowing.

Lake Victoria
Lake Victoria is the source of the upper valley of the Nile River. It lies on a recently uplifted plateau.

Kabalega Falls
The river plunges more than 130 ft (39 m) downward over this steep, clifflike waterfall, surrounded by the rising mountains of Uganda.

Rapids
A hard, rocky riverbed creates a series of rapids where the water is turbulent, tumbling in all directions as it rushes over the rocks. Rapids are often found near waterfalls.

Swamps
The Nile flows sluggishly through the reed swamps of the Sudd region. This area was once a lake.

River junction
The two main branches of the river – the White Nile and the Blue Nile – join near the city of Khartoum in Sudan.

Deep-plunge pool at base of waterfall

THE WAY TO THE SEA
The steepest part of a river's course is usually near its source in the mountains. The closer the river gets to the sea, the flatter its pathway becomes. But along the route, there may be many interruptions to this gradual change in slope. These might be lakes, where the river slowly fills in a hollow with pebbles, sand, and mud. Waterfalls form where the river runs over bands of hard rock onto softer rock.

Ancient river valleys
During the Ice Age, the climate here was wetter. Rivers from mountains near the Red Sea ran into the Nile. Today, these dried-up river valleys are called wadis.

Tributaries
The Blue Nile brings water from the Ethiopian mountains, which are hit by heavy seasonal rains. The White Nile carries water from eastern Africa. Smaller rivers that join a main river are called tributaries.

These meanders on the Darling River are cutting their way across the rocky landscape near Pooncarie in New South Wales, Australia.

MEANDERS

Where the river course runs through flat land, its path winds along in broad curves. These are called meanders. The meander curves continually get bigger and wider. This is because the water travels fastest around the outside of the curve, cutting its pathway through the riverbanks. On the inside of the curve, the water travels more slowly. Here it drops the sediment it is carrying, and forms a bank.

When it is in flood, a river can cut right through the narrow neck of land between meanders, straightening its course. The curving lake left behind is called an oxbow. An oxbow lake gradually silts up, as it is no longer part of the river.

GRAND CANYON OF THE NILE RIVER

Overall, the slope of a river's course relates to the sea level at the time. Five million years ago, the Mediterranean Sea dried up completely so that sea level, as far as the Mediterranean rivers were concerned, was 6,560 ft (2,000 m) lower than it is today. At this time, the Nile cut its pathway down to meet this deeper level. Sand and gravel carried by the river scoured away at the layers of rock, forming a steep-sided canyon.

The Nile canyon, which may have looked similar to Arizona's Grand Canyon, extended more than 620 miles (1,000 km) to Aswan. Over time, the sea level rose again, and the river dropped its sediment to fill in the canyon.

Five million years ago
The Mediterranean dried up and sea level was lower. The Nile carved a deep canyon, to try to meet the new sea level.

Present day
Later, when the sea level rose again, the canyon slowly filled in with gravel, sand, and mud from higher up the river's course.

DELTAS

Rivers in flood carry along huge amounts of gravel, sand, and mud. When a river reaches the relatively calm seawater, it drops its load of sediment in layers. These layers cannot build up thicker unless the region is sinking. Instead they get carried farther out to sea. In this way, the river may build up a triangular-shaped area of new, swampy land, crisscrossed by small channels of water. This is called a delta, after the Greek letter "delta" (Δ) which is a triangle.

A satellite view of the Mississippi River delta in Louisiana shows its bird's-foot shape.

Cataracts
Where the river passes over hard granite rocks it forms great foaming rapids of white water. These are known as cataracts. There are six cataracts on the Nile; the first, at Aswan, was the first cataract that the Nile explorers met.

Sahara Desert
The Nile finds its pathway through 1,700 miles (2,375 km) of the Sahara Desert on its journey to the Mediterranean Sea. It rains very rarely in the Sahara, so there are no tributary rivers on this part of the Nile. Nowadays, it is only fed by water from the faraway mountains upstream.

Flood waters
On the broad, low-lying plain, flood water from the Nile has spread onto the land at either side, forming a strip of green, fertile land in the midst of the brown desert.

The Nile Delta
The Nile River divides into many sluggish rivers which wander over the broad fan of the delta region in Egypt.

Bumpy bed
This is a section of a cataract. The bumpy surface of the rocky riverbed at a cataract makes the yearly flood water running over it turn a foamy white.

New land
In the delta, new land is formed by the rock and mud that the river has carried from upstream. Eventually the swampy land dries.

COASTLINES

WHERE SEA MEETS LAND, the battle between the pounding waves and solid rock creates the Earth's changing coastlines. The water of the seas and oceans is continually moving, driven by the energy of waves and currents. As the waves crash one after another against the shoreline, they find weaknesses in the rock and wear their way through. Bays, for example, are carved out where coastal rock is softer. Even cliffs made of harder rock are undercut by waves. Once undermined, the cliff breaks away in occasional rockfalls. The sea builds as well as destroys. Pieces of eroded rock are sorted out by waves – smaller ones are carried away, while larger pieces stay put, building a beach. Sand and pebbles can also add new fingerlike land to the coastline. The illustrations on these pages show how the sea is continually remodeling the coastline of Dorset, England.

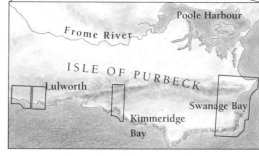

England's Isle of Purbeck
Dorset is a county in southwestern England. Its coastline stretches along the English Channel, which separates Britain from France. The area of Dorset seen on these pages is called the Isle of Purbeck.

Rocky walls
Some coasts are bordered by hard, rocky walls of granite or limestone. These walls are strong enough to resist waves and slow the rate of erosion.

Spit and tombolo
A long ridge of sand or pebbles that extends from the land into open water is called a spit. Sand is carried by drifting waves and then dropped. A tombolo is a spit between an island and land.

Hole in the roof
Waves can pound into cracks in the rock, pushing out the air inside so a spray of water rises from the rock with each wave.

Coastal rocks
The underlying rock shows why the Dorset coast has worn into an irregular shape. Headlands are found where the rock is more resistant, while bays are sculpted out of softer rock.

Layered ledges
Alternating layers of hard limestone and soft clay support a series of ledges stretching out to sea.

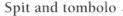

Hard band of limestone makes a vertical wall

Softer clay rocks behind the wall

Sea finds a weak spot in the wall and starts to wear through

Soft rocks are easily eroded, and are etched away by waves to form a cove

A second weak spot

Waves swirl into the cove and wear away its sides

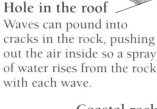

The eroded rock is deposited in the bay to form a beach

BIT BY BIT
The pounding waves can eat away the land one bit at a time. Once the waves cut through the hard wall of rock, the softer rocks behind are easily worn away. First a small cove is carved out. At high tide, the sea swirls in. The waves wear away the sides of the cove, enlarging it to make a curving bay.

Two coves may open up to form a curving bay

THE DORSET COAST
The Dorset coastline is built layer by layer of soft and hard rocks that have been gently folded. The folding has broken up some of the harder rocks, making them more easily eroded by waves pounding against the coast. Other hard rock layers form walls able to resist the waves. Soft rocks make up the low ground, which is easily invaded by the sea. Dorset's beaches are made of large hard flinty pebbles called shingle. As waves break onto the shingle their energy is absorbed, rolling the pebbles. In this way, shingle protects the cliffs and landscape from being worn away by the breaking waves.

LONGSHORE DRIFT

Each incoming wave breaks at an angle to the shore. The surf picks up pebbles and sand and moves them slightly sideways up the beach, then gravity pulls the surf straight back down to the sea. As a result, pebbles move along in a zigzag known as longshore drift. To keep the pebbles from drifting so far that they fill harbors, fences called groines are built.

Spit built by pebbles carried by the drift

Waves push pebbles and sand to the shore at an angle

Fences keep pebbles from traveling too far along

Gravity pulls the water back parallel to the shore, moving the pebbles along the beach

CHANGING SEA LEVELS

Sea levels change continuously. During the Ice Age, large changes of sea level took place – as much as 660 ft (200 m) over thousands of years. In a cold phase, sea level drops, as water from the oceans is locked up in glaciers. When the glaciers melt at the end of a cold phase, the water returns to the oceans, and sea level rises.

The Dart Estuary in Devon, England, is a ria – a river valley drowned by the sea.

A river spills into the sea
Rivers erode their valleys down to the sea at sea level. Their riverbed usually slopes very gently at the mouth, or estuary, where the river meets the sea.

The sea drowns the river
If sea level rises relative to the land, estuaries become drowned by seawater and are called rias. Rias are excellent natural harbors for ships.

Chalk hills
The chalky limestone underlying rocks makes a line of rolling hills farther inland.

A layer of chalk rock underlies this area, visible in the chalk cliffs of the coastline.

Studland Bay

Old Harry Rocks

Ballard Point

Swanage Bay

Peveril Point

Underlying layer of clay is easily eroded

Durlston Head

A dent in the land
A bay is a curving dent in the land. The waves within are usually gentle, as the headlands at either end break up the wave energy.

Sea stacks
The Old Harry Rocks were once part of the chalk headlands. But waves cut a cave into a weak spot of the rocks, eventually breaking right through it to form an arch. The top of the arch was later worn away to form sea stacks, isolated blocks of rock left stranded by erosion.

Clay beneath the bay
A bay formed here because the underlying clay rocks are softer than the chalk and limestone rocks of the headlands. The pebbly beach is made of harder rock fragments worn from the headlands.

Headlands
The resistant limestone rocks stretching out from the coast are its headlands. These shelter the rest of the coastline, protecting softer rocks farther behind. The limestone breaks up the energy of the waves, causing the water to foam in white breakers.

SAND AND SHINGLE BEACHES

Rocks worn from the coastline are ground down by the pounding waves to shingle or sand, then eventually deposited in a bay or other sheltered area to form a beach. The sand or shingle doesn't stop moving once it is laid down – in fact, beaches may sometimes change radically within hours. In a storm, large, heavy waves break directly onto the beach. These can pick up and carry the beach material into deep water offshore, so a sandy beach can disappear overnight. The beach usually returns after a long spell of calmer weather.

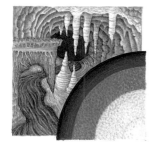

UNDERGROUND WATER

WINDING ITS WAY gently and silently through tiny spaces, water slowly travels through the rocks underground. Water is found at some depth everywhere beneath the land surface. Most of this water is rain that soaked into the ground after a shower and was not immediately used up by plant roots. Some rainwater passes through the underground rocks quickly and emerges only a few hours, days, or weeks later at a spring seeping out into a river valley. Sometimes the water stays underground for many thousands of years, either because it travels a long way or because it travels slowly. During such a long time in the rocks, water usually picks up and dissolves minerals from the rocks. In this way, the water can widen cracks in the rock until eventually an enormous underground cave forms. Underground rivers run through some caves, wearing their walls away even further. France's spectacular Gouffre Berger is shown in cross-sections on these pages.

A cross-section of the Gouffre Berger is shown here. Parts of the cave are enlarged below

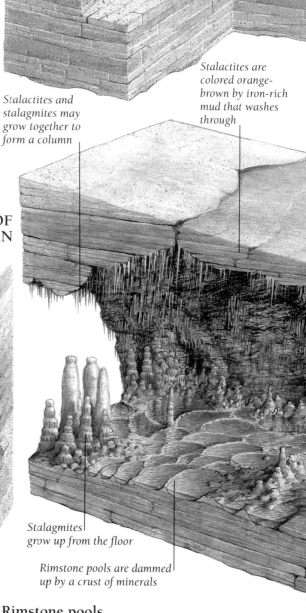

AN UNDERGROUND LABYRINTH

The Gouffre Berger (gouffre is French for "cave") is a long underground labyrinth of passageways and chambers, possibly tens of millions of years old. Its entrance is an open hole in a high limestone plateau. Today, a river runs through some parts of the cave, but others are dry. The cross-sections shown on these pages reveal some of the strangely shaped rock formations built by dripping water inside the cave. After traveling 13 miles (22 km) and dropping 3,936 ft (1,200 m), the river emerges into the light of day and joins the Furon River.

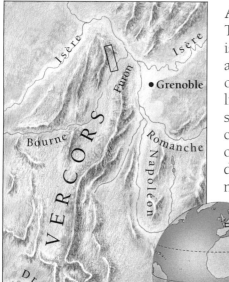

Gouffre Berger, France
The Gouffre Berger is near the city of Grenoble, beneath the Vercors Mountains. The limestone landscape here is honeycombed with caves and underground rivers.

The river emerges
During its passage through the cold, dark cave, the river picks up lime and other minerals and dissolves them. When the river emerges, the lime-rich water forms deposits called tufa as the bubbling spring water is warmed by the sunshine.

Crusts of tufa, a type of limestone

Spring water emerges from the limestone in the river valley

Stalactites and stalagmites may grow together to form a column

Stalactites are colored orange-brown by iron-rich mud that washes through

THE CANALS

Walls of the cave are hollowed out by water as it dissolves the limestone.

Underground river

Submerged river
The water level of the lakes and streams within caves changes rapidly. As soon as it rains outside, the caves fill up quickly; after a prolonged dry and hot spell, the water level may drop.

HALL OF THIRTEEN

Stalagmites grow up from the floor

Rimstone pools are dammed up by a crust of minerals

Rimstone pools
These steplike formations are made by a build-up of mineral deposits at the edge of a slope. Sometimes minerals are left behind as water runs over a slope. They form a crust, trapping a pool of water that overflows to form another step.

Cavernous cave
The biggest cave so far explored in the world is the Sarawak Chamber in Borneo, a staggering 2,300 ft (700 m) long, 1,410 ft (430 m) wide, and 395 ft (120 m) high.

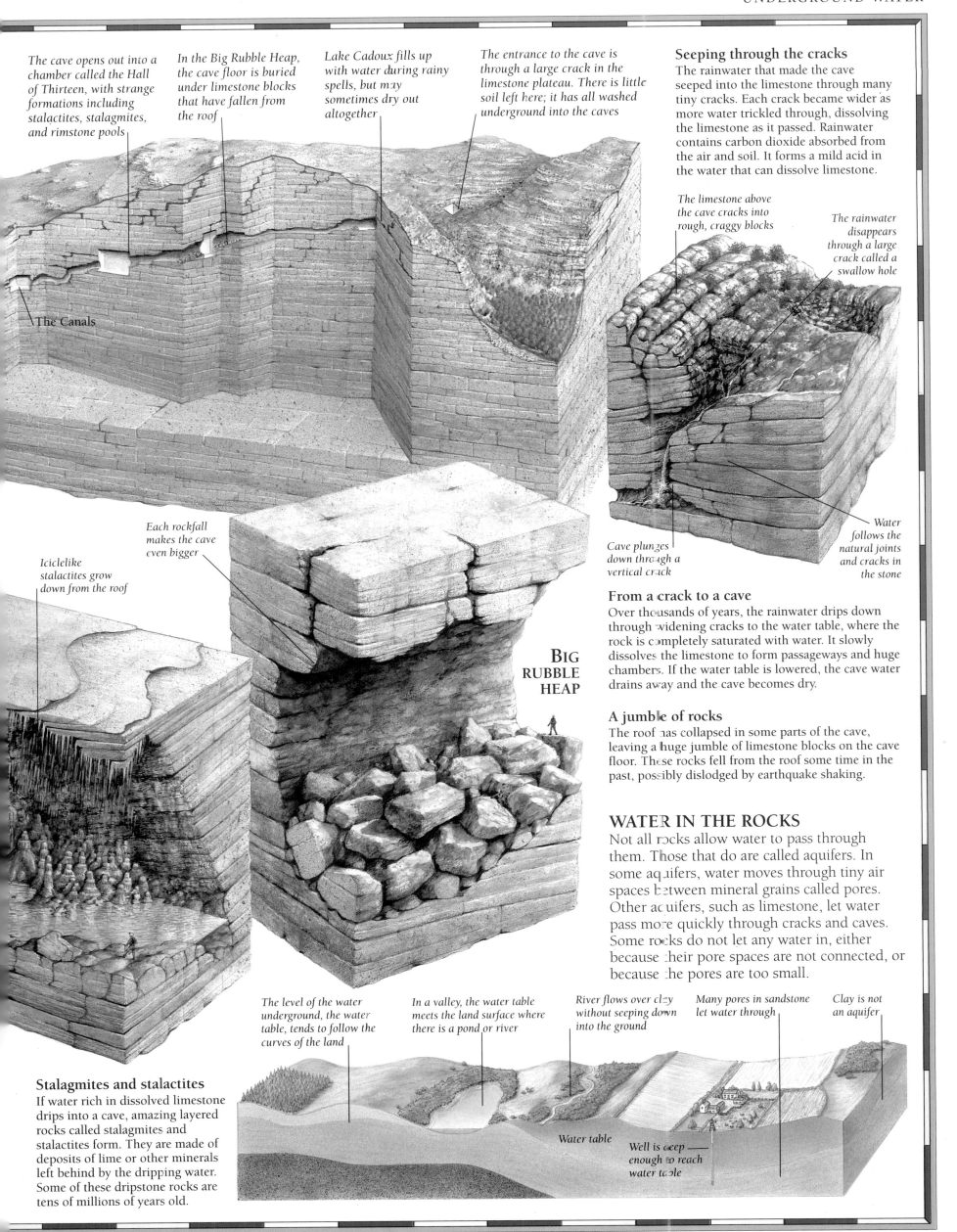

The cave opens out into a chamber called the Hall of Thirteen, with strange formations including stalactites, stalagmites, and rimstone pools

In the Big Rubble Heap, the cave floor is buried under limestone blocks that have fallen from the roof

Lake Cadoux fills up with water during rainy spells, but may sometimes dry out altogether

The entrance to the cave is through a large crack in the limestone plateau. There is little soil left here; it has all washed underground into the caves

Seeping through the cracks
The rainwater that made the cave seeped into the limestone through many tiny cracks. Each crack became wider as more water trickled through, dissolving the limestone as it passed. Rainwater contains carbon dioxide absorbed from the air and soil. It forms a mild acid in the water that can dissolve limestone.

The limestone above the cave cracks into rough, craggy blocks

The rainwater disappears through a large crack called a swallow hole

The Canals

Each rockfall makes the cave even bigger

Iciclelike stalactites grow down from the roof

Cave plunges down through a vertical crack

Water follows the natural joints and cracks in the stone

BIG RUBBLE HEAP

From a crack to a cave
Over thousands of years, the rainwater drips down through widening cracks to the water table, where the rock is completely saturated with water. It slowly dissolves the limestone to form passageways and huge chambers. If the water table is lowered, the cave water drains away and the cave becomes dry.

A jumble of rocks
The roof has collapsed in some parts of the cave, leaving a huge jumble of limestone blocks on the cave floor. These rocks fell from the roof some time in the past, possibly dislodged by earthquake shaking.

WATER IN THE ROCKS
Not all rocks allow water to pass through them. Those that do are called aquifers. In some aquifers, water moves through tiny air spaces between mineral grains called pores. Other aquifers, such as limestone, let water pass more quickly through cracks and caves. Some rocks do not let any water in, either because their pore spaces are not connected, or because the pores are too small.

The level of the water underground, the water table, tends to follow the curves of the land

In a valley, the water table meets the land surface where there is a pond or river

River flows over clay without seeping down into the ground

Many pores in sandstone let water through

Clay is not an aquifer

Stalagmites and stalactites
If water rich in dissolved limestone drips into a cave, amazing layered rocks called stalagmites and stalactites form. They are made of deposits of lime or other minerals left behind by the dripping water. Some of these dripstone rocks are tens of millions of years old.

Water table

Well is deep enough to reach water table

45

ICE REGIONS

At the ends of the earth, crowning the frozen continent of Antarctica and the icy ocean of the Arctic, are the polar ice caps. A vast amount of water is locked up inside the ice caps – they contain a hundred times more water than is in all the world's freshwater lakes put together. The water to make this ice has come from the global oceans. Ocean water is continually evaporating as warm atmospheric air circulates over it, forming moist clouds. The clouds drop their moisture as snow in cold or mountainous regions. This snow grows more compacted over the years and turns into glacier ice. In order for glaciers and ice caps to grow larger, the rate at which the snow falls must be much greater than the rate at which it melts, so that there is snow left over at the end of each summer. During the great Pleistocene Ice Age over the past two million years, much of North America, and the whole Baltic Sea, were under ice caps. At times when there was so much ice, global sea level was much lower.

Storm waves have eroded this Arctic iceberg into pinnacles. Icebergs are not made of frozen seawater. They are broken off from the end of an ice sheet or glacier, so are composed of frozen fresh water.

THE FROZEN CONTINENT

A vast ice sheet covers much of the frozen continent of Antarctica. It has been accumulating for tens of thousands of years, and today it is over 14,800 ft (4,500 m) thick. The ice moves slowly outward and downhill, and toward the frigid seas. On the coastline, the ice is thinner, so that the summits of high mountains peek through. The sheet of ice extends out from the continental landmass, floating on the sea to form an ice shelf. At its edge great lumps of ice split off from the main mass and float away as icebergs – "berg" is a German word which means "mountain." Three sections of Antarctica are illustrated here.

Floating mountains of ice

Only a small part of an iceberg floats above the sea. Most of its mass is below sea level. The ice contains rocky boulders scraped off Antarctica as the ice sheet moved toward the coast. When large icebergs break off, or calve, from the ice sheet, they make huge waves.

Glaciers weave their way through the nunatak mountain tops

Meteorites that fell thousands of years ago are revealed in the glacier ice

Floating sea ice

The edge of the Antarctic ice sheet spreads out as shelves of sea ice. It floats because it is less dense than the ocean water, which is salty as well as very cold.

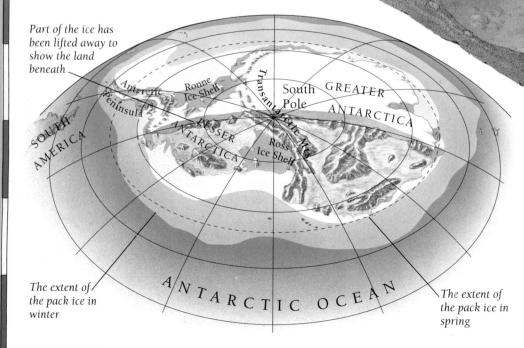

Mountains peeking through the glacier ice are called nunataks

Part of the ice has been lifted away to show the land beneath

The extent of the pack ice in winter

The extent of the pack ice in spring

Seabed deposits

The seabed is covered with layers of clay and sandy gravel dotted with boulders, all deposited here from melting of the icebergs.

THE ICY LAND OF ANTARCTICA

This vast continent is about one and a half times the size of the US. Ice covers all but about 5 percent of the land. Although Antarctica seems to have been near the South Pole for the past 200 million years, it was not always a frozen land. Around 35 million years ago, it became totally separated from the other continents as Pangaea split up. Perhaps it was then that the climate began to change. Once the cold Antarctic Ocean currents could circulate all the way around Antarctica, it was isolated from warm tropical ocean currents. This may have been enough to trigger heavy snowfall, which led to the growth of ice sheets.

THE ICY WATERS OF THE ARCTIC

There is no continent at the North Pole. Instead, the region includes the Arctic Ocean, the northernmost parts of North America, Europe, Asia, Greenland, and several smaller islands. The surface of the ocean is covered with salty sea ice, formed from frozen seawater. Pack ice, broken and crushed together again by the movement of the water, fringes the sea ice. The extent of the pack ice varies from season to season – about half of it melts in the summer. Glaciers crisscross the land in and near the Arctic. But in Siberia, some parts of Alaska, and Canada, there is too little snowfall for glaciers to form. The winter air is so cold that the water in the ground freezes to such a depth that it never completely thaws. This frozen ground is called permafrost.

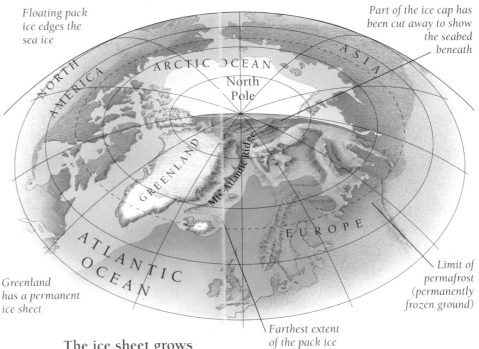

Floating pack ice edges the sea ice

Part of the ice cap has been cut away to show the seabed beneath

Greenland has a permanent ice sheet

Limit of permafrost (permanently frozen ground)

Farthest extent of the pack ice

Glacier ice has scraped the mountains into rugged peaks

At the South Pole, the ice sheet is 9,000 ft (2,800 m) thick. The rocky land beneath is at about sea level.

The ice sheet grows

As the ice sheet grew over Scandinavia, it spread out. But it also became thicker, so the surface of the ice grew higher. The ice also added to the weight of the land mass, which then pressed down on the squishy asthenosphere. The ice grew thicker and higher much faster than the asthenosphere was able to sink lower out of the way.

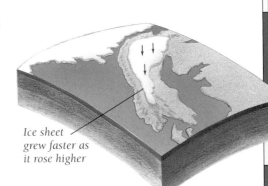

Ice sheet grew faster as it rose higher

Full press

As the ice sheet reached its full extent, at last the asthenosphere caught up and flowed away. The land level became lower. It became so low that the ice sheet was no longer high and cool enough to collect snow, and so it began to melt and shrink.

Asthenosphere material flows away sideways

Thick ice depresses the whole land mass

Frozen summits

On its way toward the sea, the ice sheet must flow around the peaks of the mountains. Cold, dry winds whip down off the high central ice plateau, whistling through the passes between the mountain summits. Because the winds are dry, they strip off and erode the upper layers of ice, revealing meteorites that fell onto the ice thousands of years ago.

Under pressure

The top of the thick ice covering Antarctica is well above sea level. But in the vast interior of the continent, the weight of the ice has pushed the rocky land below sea level. Some of the deep ice may be very old, formed by snow that fell up to a million years ago. Just as rocks recrystallize under pressure, the deep ice slowly recrystallizes under the massive weight of the overlying ice.

Lifting land

Today, the asthenosphere is still slowly flowing back, long after the ice sheet has melted away. The plate under the Baltic Sea may eventually bounce back so much that the sea becomes dry land. Many shorelines show evidence of huge isostatic uplift over recent centuries.

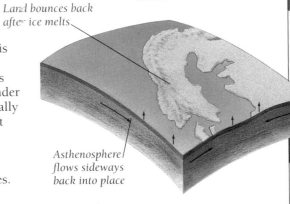

Land bounces back after ice melts

Asthenosphere flows sideways back into place

DEPRESSED BY ICE

Even though ice is not as dense as rock, the weight of an ice sheet covering a landmass is considerable. It alters the balance of the plate, which floats on the squishy asthenosphere. As the ice sheet builds up, the asthenosphere flows out of the way, and the plate sinks. If the ice melts, the asthenosphere slowly flows back and the plate rises. These changes in land level are called isostatic changes. The sequence above shows isostatic changes created by an ice sheet over Scandinavia 30,000 years ago.

Antarctica's Ross Sea (above) is fringed by an ice shelf, and is littered with icebergs calved off from the main ice sheet.

This is a raised beach on the Isle of Mull, Scotland. The land has risen, and the old shoreline is now left high and dry.

RIVERS OF ICE

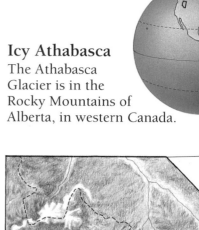

GLACIERS SNAKE DOWN THE VALLEYS of many of the world's mountain ranges. These huge masses of ice are made from layers of snow that build up until the increasing weight causes solid blocks of ice to form. The ice becomes so thick and heavy that it starts to move – either outward in all directions, such as the vast domelike ice sheets of Greenland and Antarctica, or down a valley, such as the Athabasca Glacier seen here. Glacier ice travels slowly, less than 3 feet (1 meter) a day. Grinding against the surrounding valley sides, the glacier scrapes away any loosened rocks, which are carried downhill with the glacier ice. Eventually, far down the valley where the glacier melts, rocks, boulders, and fine rock dust are dumped in great heaps. At the moment, most glaciers are melting faster than new ice is forming in the high mountains. This means that glaciers are getting smaller. Glaciers have been retreating for 10,000 years.

Icy Athabasca
The Athabasca Glacier is in the Rocky Mountains of Alberta, in western Canada.

A river of melted ice runs around rocks at the snout of a glacier in Norway. The rocks have been polished smooth by the action of the sandpaper-like rock dust that is frozen within the glacier.

THE ATHABASCA GLACIER
The glacier flows out from an ice sheet, the Columbia Icefield, in the snowy mountains around the 12,293-ft (3,747-m-) -high Mount Columbia. The icefield was once much larger. At that time, the glaciers in the side valleys were all part of the main glacier. Today, the ice level has dropped so low that they no longer meet the main glacier. Instead, they hang isolated above the main valley. The end of the glacier is a heap of boulders and rock dust, called the terminal moraine. The glacier meltwater is the source of the Athabasca River, which flows into Hudson Bay and the Arctic Ocean, thousands of miles (kilometers) away.

Hanging valley
This valley, carved by a glacier long ago, now "hangs" above and to the side of the main valley.

The snout, or end, of the glacier is colored gray from all the rock fragments and dust in the ice

Moraine left by an earlier glacier

Glacial gravel
When glaciers advance over soft sediment, such as clay or gravel left by a previous glacier, they smooth it into long, egg-shaped hills. These streamlined hills, called drumlins, may be up to 164 ft (50 m) high and 6,562 ft (2,000 m) long. Drumlins are usually in groups, forming what is known as a "basket of eggs" landscape.

Long side of a drumlin is parallel to the ice flow

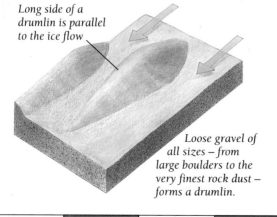

Loose gravel of all sizes – from large boulders to the very finest rock dust – forms a drumlin.

Rock ridges
These ridges of rock deposits are old terminal moraines, from when the glacier was longer and thicker. The gravel is now being eroded and carried away by the rivers of meltwater.

Milky rivers
Meltwater pours out from ice caves under the glacier to feed rivers. The water is milky green because it contains fine rock dust.

Lake filled with glacier meltwater is dammed up by the gravels of the terminal moraine.

Ice on the move

When the glacier ice becomes thick enough, it starts to move under the pressure of its own weight.

Crevasses

Deep crevasses, or cracks, open up on the surface of the glacier as it travels over steep or rugged terrain.

Heavy winter snowfall "tops off" the ice dome that feeds the glacier.

Steep slope

The slope at the head of a glacier is steep because the glacier has ground rock off the mountainside.

Glued to a glacier

Melting water trickles down the rock face behind the glacier, seeping into cracks in the rock. Under the glacier, the water freezes again, "glueing" the rock to the glacier so it is dragged away.

Back wall of mountains stays rocky and steep.

Melting water trickles behind the glacier.

Ice sheet

The ice sheet spreads out sideways, the glacier ice licking its way down through the mountain peaks.

Inside the glacier

The ice here is dirty white with all the rock fragments it has picked up. The fragments grate against the valley bed.

Moraines

Rock fragments are frozen into the glacier ice in long bands called moraines. It takes thousands of years for one rock from a mountain peak to travel all the way to the glacier snout, where it is dropped in a long ridge known as the terminal (end) moraine.

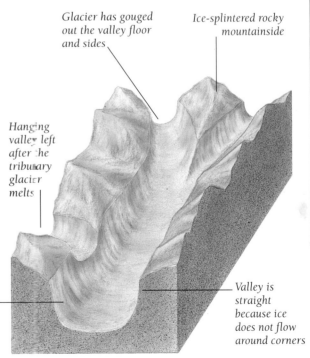

As two glaciers meet in the mountains of Switzerland, their lateral (side) moraines converge to make a dark stripe down the middle, the medial moraine. Where several glaciers meet, there are many of these stripes.

Glacier lifted away to show how the valley beneath has been widened and gouged out to a broad U-shape.

Icebreaker

The mountain surface is eroded by melting snow running into cracks in the rocks. At night it freezes to ice, expanding and breaking apart the rock.

Polished smooth

The valley walls and floor are rasped smooth by the rocky ice, sometimes to a fine polished surface.

GLACIER VALLEYS

A glacier makes its valley wider and deeper as it carves away at the mountainsides. It scoops everything from rock fragments to huge boulders from the mountain by ice-plucking. This happens when water that seeps into cracks in the rock freezes, attaching the rock to the glacier. The resulting rock-laden ice scours the valley walls and floor as it grinds over them. Rock fragments dislodged from the steep, rocky mountain cliffs fall to the sides of the glacier and are carried away.

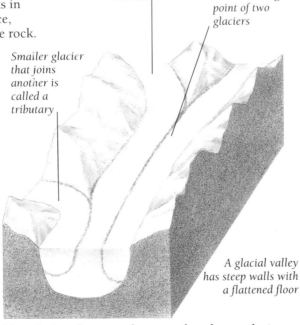

Valley filled with ice

Medial moraine at the meeting point of two glaciers

Smaller glacier that joins another is called a tributary

Glacier has gouged out the valley floor and sides

Ice-splintered rocky mountainside

Hanging valley left after the tributary glacier melts

A glacial valley has steep walls with a flattened floor

Valley is straight because ice does not flow around corners

Two glaciers flow together to make a larger glacier that fills the valley with ice. A smaller glacier from a valley to one side merges into the large glacier.

After a glacier has melted, its straight U-shaped valley is revealed. Tributary valleys hang higher above the level of the main valley.

DESERTS

AMONG THE MOST DESOLATE places on Earth are its deserts – deserted areas where almost nothing lives. Some deserts are hot and dry all year round, others are dry with intensely cold winters. The cold lands of the Arctic and Antarctic are also deserts. Deserts have thin soil or none at all, so there is little vegetation. Instead, the land is littered with stones or covered with bare rock or shifting sand dunes. The sparse plant life means few creatures can survive. Deserts have very low rainfall. Only a few inches (centimeters) might fall a year, though years may go by when no rain falls at all. The dry air over the desert means that any surface water rapidly evaporates. During the rare rainstorms, there is no vegetation to slow down the running surface water. Flash floods create lakes, which then quickly dry up leaving behind salt flats. The Earth's desert regions are shown in this map.

In the tundras of the northernmost part of North America, almost all the water is frozen beneath the surface.

NORTH AMERICA

Because mountains near the Pacific coast catch all the rainfall, the air inland is dry.

— Death Valley

Sonoran

Tropic of Cancer

PACIFIC OCEAN

ATLANTIC

Equator

SOUTH AMERICA

Cold ocean currents off South America cool the wind blowing on land. Any moisture makes fog at the coast – so very little reaches the Atacama Desert.

Tropic of Capricorn

Atacama

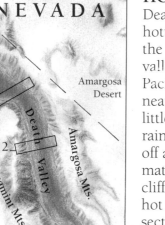

NEVADA

Mesquite Flats

Section 1

Amargosa Desert

Death Valley

Section 2

Amargosa Mts.

Panamint Mts.

CALIFORNIA

HOT, DRY, AND LOW

Death Valley tops the lists of places with the hottest temperatures, the lightest rainfall, and the lowest elevation in North America. This inland valley is a desert because the moist winds from the Pacific Ocean drop all their rain on the mountains nearer to the coastline. There is no soil and very little vegetation. On the rare occasions when it rains, the water pours over barren rock, scouring off all loose fragments and sand. The coarse rocky material is dropped at the foot of the mountain cliffs, while sand and salty silt is spread out on the hot valley floor, where it eventually dries out. Two sections of Death Valley are illustrated below.

This is one of the many huge playas, or dried-up lakebeds, in the Atacama Desert in Chile, probably the driest place on Earth.

California's Death Valley
Death Valley is one of hundreds of salty land-locked valleys in the western states of California, Utah, and Nevada.

Small windblown fragments of loose rock collect as sand dunes

When it rains, the water fills a lake which dries in the hot sun

— Thimble Peak

Mount Perry

Rugged mountain slopes where no plants live

Mesquite Flats

Dante's View

Death Valley Canyon

Badwater Basin

Wadi, or dry gully, becomes a raging torrent of muddy water when rain falls.

Hanaupah Canyon

Lowest elevation in North America, 282 ft (86 m) below sea level

The valley formed when blocks of crust sank along the fault line

Water carries loose sediment down the wadis to collect in fans at the base of the canyon.

This line represents sea level

Fault along which the two sides of the valley have separated

Salt flats, where salt crystallizes out of the lake water

Salt flats
Although there is rarely enough water in a desert to make a lake, rainstorms and flash floods fill a temporary lake called a playa or salina. When the playa dries out, the salts dissolved in the water recrystallize, leaving a gleaming white layer of salt covering the flat valley floor. The salt flat grows thicker each time the playa fills and dries out again.

The Sahara (the world's largest desert) and Arabian deserts are hot and dry tropical regions, north of the Equator.

The Gobi and Taklimakan deserts are in the middle of a huge continent, far from any moist winds.

When the Himalayas were pushed up, they stopped moist air from reaching the interior of Asia. This is how the Gobi Desert area became so dry.

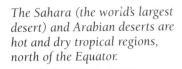

EUROPE · ASIA · Taklimakan · Gobi · Sind · Sahara · Arabian · AFRICA · Rub Al Kali · PACIFIC OCEAN · OCEAN · INDIAN OCEAN · Namib · Kalahari · AUSTRALIA · Gibson · Simpson · ANTARCTIC OCEAN · ANTARCTICA

The Namib Desert, like the Atacama in South America, is dry because of cold ocean currents nearby.

Almost two-thirds of Australia is a desert, at its driest in the flat, barren plateaus of the center.

Underneath the sands of the Kalahari Desert is a plateau built from lava flows 65 million years ago.

Violent winds

The wind whips up dust storms which blow through deserts, removing fine silt and sand, and dumping it elsewhere. They are also known by the Arabic name haboobs, which means "violent wind."

Tropical deserts

Although seasonal rains do fall in tropical deserts, sometimes in vast downpours, these areas are so hot that they quickly lose the moisture through evaporation.

KEY TO MAP

- Extremely arid desert regions, where there is no rain for an entire year
- Arid desert regions, where any rainfall evaporates quickly
- Semi-arid desert regions, where under 20 in (50 cm) of rain falls a year
- Frozen tundra desert regions, where cold dry air means there is little rain

Frozen deserts

Although an enormous amount of water exists in the frozen regions of Greenland and Antarctica, these places are still deserts, because it is so cold that the water is all frozen into ice. It sometimes rains in Greenland, but only snow falls in Antarctica – and very little. Some places here are so dry that ice evaporates in the dry wind.

Desert dunes

When most people think of deserts, they imagine sand dunes, hills of loose sand deposited by the wind. In fact, only about 20 percent of the world's deserts are sandy. In some of these deserts, the underlying rock is covered by only a few inches (centimeters) of sand, but in others there is enough to pile into huge dunes up to 1,640 ft (500 m) tall. Even a small pile of sand is big enough to block and trap other grains drifting by, so that the dune grows larger.

These towering sand dunes rise above the Sahara Desert in Algeria.

SAND DUNES

The shape of sand dunes depends on the strength and direction of the wind. Crescent-shaped barchan dunes form where wind moves the low crescent ends faster than the high middle part of the dunes. Ridge-like seif dunes are found where the wind direction is strong and steady.

Barchan dunes

The crescent ends point downwind.

These dunes form where the desert floor is hard and flat, and wind blows regularly in the same direction.

Seif dunes

Dunes are in straight lines.

The wind spirals along, blowing sand to the sides to form long ridges.

How dunes move

Sand dunes move about 82 ft (25 m) a year. Wind blows sand up the gentle slope of a dune, and the steep slope becomes steeper and steeper until it collapses in a small avalanche. The sand comes to rest in sloping layers parallel to the avalanche tracks, called cross-bedding.

Dune moves in this direction, blown by the wind

Older dunes are covered up by newer ones

Sand of old dunes is compacted and their structure is preserved

SOIL SUPPORTS LIFE

WITHOUT SOIL, there would be little life on Earth. Soil is the link between life and the rocky part of the Earth, supporting the plants that nourish people and animals. Soil is made where rocks are weathering and softening, breaking up into smaller particles, and giving up their rich store of chemicals in a form which plants can use. Soil contains mineral grains, air, and water. It is also rich in organic material, such as plant roots, fungi, beetles, and worms, as well as microorganisms such as bacteria and algae. The water and air in the soil are vital to plants. If soil becomes waterlogged, plants suffocate, and the soil becomes putrid with undecayed plant and animal remains. If there is no water, plants wilt and die and the dry soil may disappear on the wind. It takes a long time for a thick, nourishing soil to develop over the underlying rock.

Hillsides of the world
The soil profiles below are found in three different climates: temperate, arid, and tropical. The thickness and richness of soil depends on climate and many other factors: underlying rock type, the age of the soil, the landscape, the drainage, the vegetation, and the diversity of animals living in the soil and nearby.

Temperate climate soil
In a temperate climate, the soil is thicker in the valley than on the hillslopes, because rain and frost – and gravity – move soil downhill.

Roots penetrate into soil and make channels for air and water. These tree roots also help prevent soil from being washed away by heavy rain.

SOIL PROFILE
By digging down into the soil right through to the rock below, it is possible to see the entire profile of soil. Soils are made up of layers known as horizons. Each horizon has its own unique physical, chemical, and biological characteristics. The topsoil layer at the surface is rich in organic material. Below this is the subsoil, penetrated by a few roots. Deeper still is a layer of weathered rock fragments and boulders worn from the solid bedrock.

A closer look at soil
This is a section of soil viewed through a microscope. It shows the dead and decaying plant and animal material that will eventually nourish living plants growing in the soil. Air and water fill the spaces between the decaying material, along with microorganisms and tiny plants.

Mat of plant roots in topsoil

Rabbit burrow

Topsoil
The dark, rich topsoil is matted together with plant and grass roots. Topsoil contains humus – the remains of plants and animals in the process of being broken down to simpler chemicals. Bacteria and fungi within the topsoil help this to happen.

Animal burrows, whether made by a tiny earthworm or a large rabbit or badger, let air into the soil and allow water to drain through

Subsoil
There is less organic material in the subsoil. This horizon is rich in mineral particles weathered from the solid rock beneath. These minerals contribute new plant nutrients to enrich the soil above.

Underlying rock
Bedrock is at the bottom of the soil profile. The nature of the bedrock is important to the composition of the soil above. Some rocks contain more chemicals that are useful to plants as nutrients, and therefore make richer soils.

Air and water fill any gaps in the soil

Woody tissue being broken down into humus

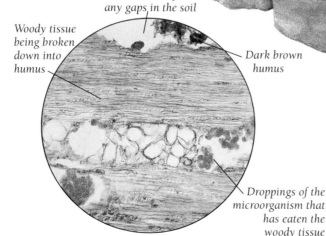

Dark brown humus

Droppings of the microorganism that has eaten the woody tissue

Left behind
Animals eat grass and browse the shrubs, but in turn, they leave behind their droppings, which fertilize the soil. Fallen leaves and twigs are also broken down into humus.

Humus and clay particles have large spongy surfaces, that can hold and exchange plant nutrients

Bedrock
The constant weathering of the bedrock below the soil helps to make the soil thicker. The basic texture of the soil depends to a great extent on the type of bedrock underlying it.

SOILS ON SLOPES
Soil continually moves downhill. When hillslopes are plowed or are bare of vegetation in winter, heavy rain may wash soil into the valley below. Animals may also dislodge soil and push it slowly downhill as they walk over, or burrow through, steep slopes. With each frost, ice lifts soil particles out from the hill, then drops them lower down in the valley.

Arid climate soil

The sparse vegetation in an arid (very dry) climate cannot prevent occasional rainstorms from washing the soil away downhill. When it is dry, the wind blows the soil away. As a result, the soil is thin.

There is no soil on a rocky hilltop. Instead, the rocks are split by temperature changes between day and night, winter

Tropical climate soil

Thick, quick-growing forest shrubs and roots of trees hold the soils in place, so hillsides as well as valleys in tropical climates have thick soil.

Tropical vegetation grows quickly, and it supports many kinds of animals

In a warm, moist tropical climate, bacteria and fungi work fast to make a thick blanket of humus

A part of the southwestern U.S. known as the Dust Bowl was devastated by dust storms in the 1930s. They buried farms and roads and made farming impossible. Because the soil had been plowed up and the crops taken off, there were no longer roots to hold the soil in place. It became dry, and blew away with the wind.

Rock weathers rapidly in a tropical climate, enriching the soil with plant nutrients

How tropical soil weathers

The damp climate and rapid rock weathering means the soil becomes thicker with time. In spite of heavy rain, the soil is not washed away as long as plant roots hold it in place. When they die, plants decay and return their nutrients to the soil.

Because there is little moisture to support vegetation, there is little humus in the arid soil

Hot, wet climate and strong sunshine allow quick plant growth

As plants die, they decay and add to the soil

Fast-weathering rocks also add to the soil thickness

Dry mineral grains are blown away because there are few roots to anchor soil

Slow weathering in a dry climate cannot keep pace with the soil blown away

Rocky surface left behind has little soil

How arid soil weathers

Arid soil has few plants and roots to hold it in place. The mineral grains are easily blown away. Because weathering is so slow in an arid climate, the soil tends to be shallower and less developed than soils in warm, moist regions.

In the valley

The moist climate and continual weathering of the underlying rocks helps maintain the rich humus in the valley, making a nourishing soil for a variety of plants.

HOW SOIL FORMS

The process of soil formation in a cold climate is shown here. Glaciers strip away all soils over which they pass. When they melt, new soil forms slowly from rock weathering. At first, there is little in the way of plant and animal life to help make humus. But mosses and scrubby bushes soon grow, and the process of breaking down the bedrock begins.

Glacier strips away the soil

Bare rock and gravel remains

Moss and scrubby bushes grow

Small trees gain a foothold and start to contribute their decaying leaves to new soil

More animals move in, and their droppings enrich the soil

The topsoil horizon grows thicker as the underlying rock weathers away

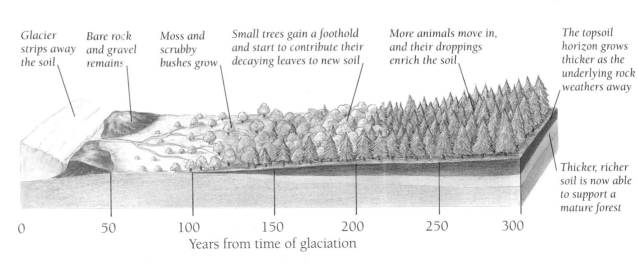

Thicker, richer soil is now able to support a mature forest

| 0 | 50 | 100 | 150 | 200 | 250 | 300 |

Years from time of glaciation

EARTH'S INGREDIENTS

MILLIONS OF STARS populating the universe long ago created the ingredients of our Earth. Each star was a factory for chemical elements, turning hydrogen and helium into other, heavier elements. Some stars exploded at the end of their lives, scattering their materials out through the universe. Then 5 billion years ago, some of this stardust clumped together to make Earth and the rest of the solar system. Since its formation, the planet Earth has shaped and regrouped the chemical elements that make it up. Its solid surface is made up of what at first seems like a bewildering variety of rocks. In fact, only a small number of different minerals make up most Earth rocks. Less than ten of the hundred or so chemical elements known on Earth are found in these rock-forming minerals. The tiny atoms that combine chemically to make up the Earth's minerals are made of even smaller particles, called protons, electrons, and neutrons.

Swirling white cloud patterns in the atmosphere

The solid rocky surface is seen in yellow and green

White cloud appears where the atmosphere contains masses of water droplets

The oceans are shown in blue

Earth and its neighbors
The Earth is one of nine planets that travel around our local star, the sun, in more or less circular orbits, or paths. Only Mercury and Venus are nearer to the fiery sun.

Colliding in space
As the planets spin, they collide with smaller fragments. Most of these fragments were swept up in the first billion years of the solar system's history. Some remain as asteroids, and others crash onto planets as meteorites.

All life on Earth depends on the light and heat of the sun

The nine planets and about 70 moons condensed from the same dust cloud as the sun

The solar system
More than 5 billion years ago, somewhere toward the end of one of the spiral arms of the Milky Way galaxy, a dust cloud began to gather. It was mostly made of hydrogen and helium, but also contained heavier elements created in previous supernova explosions. As the cloud became hotter and thicker, gravity pulled a clump of material toward the center, while the rest of the gas and dust flattened into a spinning disk. The central clump condensed to become the young sun. Smaller clumps of matter still spinning around the center became the planets of the solar system.

An exploding supernova shines brighter than a billion suns put together

BIRTH AND DEATH OF STARS
The most common chemical elements in the universe are hydrogen and helium. All the other chemical elements are made within stars. Stars have a long but definite life cycle. After a star is born, hydrogen is converted to helium in its core, as the star glows white and bright with light and heat energy. When there is no more hydrogen left, the star begins to die. Some massive stars explode as supernovae. The huge collapse which begins the death throes of the star also raises its temperature, and brings about a synthesis of new, heavier chemical elements. Seconds later an explosion follows, which blasts these new chemical elements out across the universe. In this way, the whole mix of about a hundred chemical elements are made.

PLANET EARTH

Earth's surface is unlike that of any of the other planets. Earth is surrounded by an atmosphere of gas, and nearly three-quarters of its solid surface is hidden under water. This gas atmosphere and liquid hydrosphere have separated from the solid part of the planet over thousands of millions of years. The most common element, or basic chemical, in Earth's crust is oxygen. Although oxygen is usually a gas, it is combined in the planet's rocks with the element silicon. Because the two bind strongly together, the silicate rocks they form are chemically stable.

The only planet with liquid water is the Earth – other planets are too hot or too cold

Earth's atmosphere protects life on the surface from some of the harmful rays of the sun

TABLE OF ELEMENTS

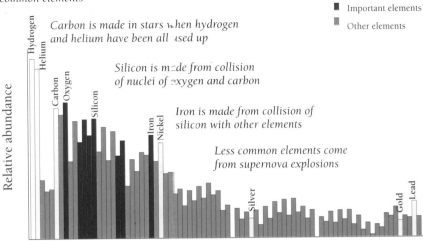

Hydrogen and helium are the most common elements

Carbon is made in stars when hydrogen and helium have been all used up

Silicon is made from collision of nuclei of oxygen and carbon

Iron is made from collision of silicon with other elements

Less common elements come from supernova explosions

Key to table
☐ Rock-forming elements
■ Important elements
▬ Other elements

Relative abundance

Increasing atomic number (number of protons)

An abundance of elements

Chemical elements are basic substances that make up minerals. The table above shows the elements in the universe – so far, 109 are known, including 89 found naturally on Earth. Some elements are much more common than others. Hydrogen is the simplest and most fundamental element from which all others are made. In the intense heat of a star, for example, hydrogen nuclei collide to make helium. When all the hydrogen is used up, the star uses its helium nuclei to make carbon and oxygen, its carbon to make magnesium, then oxygen to form silicon, and finally silicon to make iron. Other, less common elements are only made in supernova explosions.

Serpentinite rock is made from crystals of olivine and pyroxene which have been changed by adding water into their crystal structure

Rock ingredients

Minerals, naturally formed solids with crystalline structures, are made up of elements. Almost all rocks are made of silicate minerals, formed by a strong bond between the elements oxygen and silicon. Other common elements are iron, magnesium, aluminum, calcium, potassium, and sodium. Together, these eight elements make up most of the rocks in the Earth's crust. This serpentinite rock is especially rich in iron and magnesium, which make it dark in color.

Gray striped crystals are feldspar minerals

Bright colors of the mineral olivine

Mineral mixtures

There are several thousand different kinds of minerals, but only a few dozen are common. Rocks are usually made of a mixture of three or four common minerals, with a scattering of more unusual ones. The different amounts and different types of minerals present give the great variety of rocks that cover the Earth's surface – from soft sand and clay to granite, limestone, and basalt. This microscopic view shows a slice of gabbro, a rock made of three main minerals, feldspar, olivine, and pyroxene, with smaller amounts of iron oxide.

Black minerals are iron oxide

Under the microscope

The minerals that make up rocks can be seen as individual crystals when viewed through a microscope. Most minerals are transparent when sliced thinly enough.

Twenty-six electrons whiz around the nucleus

Structure of atoms

The smallest particle of a chemical element is an atom. Even smaller particles called protons (which have positive electrical charges) and neutrons (which are neutral) form the nucleus, or core, of an atom. Electrons (which have negative charges) whiz around in a cloud outside the nucleus. Each chemical element has a different number of each kind of particle. The number of protons and electrons controls which chemical elements combine with what others in order to make minerals. The atomic structures of three elements – hydrogen, carbon, and iron – are illustrated here.

Nucleus of one proton

Nucleus made of six protons and six neutrons

Six electrons round the nucleus

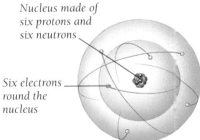

Hydrogen

Hydrogen atoms, the most abundant element in the universe, are also the simplest, with just one proton circled by one electron. Hydrogen combines with other elements to form compounds, such as water.

Carbon

This element is made in the dying stages of a star's life. The carbon in humans, trees, and rocks all came originally from stars.

Iron

In the whole Earth, iron is the most abundant element, but most of it is concentrated in its heavy metal core. Iron is common in many minerals, and it is the chemical element that adds color to most rocks. It is the heaviest element made when ordinary stars die.

IGNEOUS ROCKS

WHEN PLANET EARTH became cool enough to have a solid outer skin, igneous rocks were the first to appear on its surface. Igneous rocks are still being made today as volcanoes erupt molten magma, which cools and becomes solid. Some magma, though, does not get to the surface. Instead, it cools as igneous rock underground. Many millions of years later, the igneous rocks that cooled several miles (kilometers) down in the crust may become exposed at the Earth's surface by uplift and erosion. There are many different types of igneous rocks that crystallize from different magmas under different conditions. They are grouped by their textures and by the minerals they contain.

Basalt with amygdales
Gas bubbles in lava fill up with minerals to make amygdales.

Pegmatite
Pegmatite contains unusually large crystals.

Mica
The shiny dark flakes seen in some granites are mica. In granites with large crystals, bits of mica can be flaked off in thin sheets.

GRANITE

If it were possible to put all the rocks of a continent in some kind of giant crushing machine, mix up the crushed rock, melt it, and then let it cool and crystallize, the result would be granite. Granite is one of the most common igneous rocks, found at the core of many mountain ranges. The first continents were made of granite-like rock. Granite contains the minerals quartz and feldspar, along with a small amount of dark minerals, such as mica. Their crystals can be seen in the granite on the left: quartz is gray, feldspar is pink or white, and mica is black.

Quartz
Glassy quartz, an essential mineral in granite, may be transparent, but is sometimes a milky blue.

Feldspar
Igneous rocks are classified by the amount and type of feldspar they contain. It is one of the most common minerals in the Earth's crust.

Granite

Gabbro
Gabbro crystallized slowly like granite, but has more dark minerals and usually no quartz.

Agate
The beautiful bands of agate are formed, one layer after another, as they coat the inside of gas bubbles in volcanic rock. The outermost agate layer forms first, then in time, each layer afterward forms toward the center. Sometimes there are quartz crystals lining the inner core of the agate.

Gold
Some chemical elements, such as gold, do not crystallize easily as magma cools and hardens. These awkward elements get concentrated in the last bit of magma liquid, and finally crystallize in cracks called veins, which open up as granite cools and shrinks.

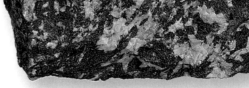

Silver
Silver, a shiny, gray-white metal, is one of the few metals that crystallizes from magma without combining with other chemical elements. These are called native metals; they need almost no processing before they are used. Other native metals include gold, platinum, and copper.

Diamond
One of the most amazing minerals is diamond, the hardest natural substance known. Its dense structure is a result of crystallizing under great pressure. Diamonds are brought to the Earth's surface by volcanic eruptions.

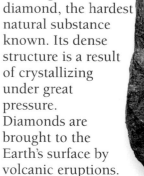

A diamond crystallizes about 62 miles (100 km) below the surface

LAVAS

Some of the rocks formed by recrystallized lava are shown here. The texture of the rocks depends on the type of lava. Some lavas are very hot when they are erupted, spreading out quickly as they cool. Others are cooler, and move only slowly, hindered by the crystals as they shape and grow.

Vesicular basalt
Volcanic gas bubbles out of magma and may be trapped as the lava solidifies. The bubbles in this rock are called vesicles.

Obsidian
This lava erupts at a low temperature, with its chemical framework already in place. This "freezes" to obsidian, which has no crystals.

Pumice
This is a glassy lava froth, and is often part of the same eruptions as obsidian. Some pumice is so frothy and bubbly that it floats on water.

Volcanic bomb
Hot basalt lava cools so fast that blobs like this, which are thrown out in explosions, have solidified by the time they land on the ground.

This is a crystal of augite, one of the minerals in the pyroxene family, embedded in an igneous rock

BASALT

Most of the solid surface of the Earth is made of basalt. All the solid ocean crust is basalt, and there are also many huge basalt lava flows on the continents. Basalt is formed by molten rock in the Earth's mantle. It is fine-grained and dark in color because of the dark minerals it contains: principally pyroxene and olivine. If basalt magma cools slowly, it grows larger crystals. Then it is called dolerite, or if the crystals are really large, gabbro.

Basalt

Pyroxene
Pyroxene is a group of dark, dense minerals that makes up basalt, gabbro, and dolerite. It is rich in the chemical elements iron and magnesium.

Feldspar
The dark, heavy rocks on this page contain calcium-rich feldspar. The lighter granites contain sodium- and potassium-rich feldspar.

Olivine
Olivine is a shiny green, heavy mineral, rich in iron and magnesium.

Ropy lava
The twists and turns in this rock are formed when hot flowing lava causes wrinkles in its cooler crust.

Sulfur is scraped from volcano craters. The workers who collect it risk their lives every day

Sulfur
These lemon yellow crystals are found around many volcanic craters and hot springs. Sometimes sulfur is combined with other chemicals to make sulfates or sulfides.

Copper is flexible, making it useful for bendable pipes

Hematite
Hematite is an oxide of iron – iron gone rusty. Volcanic gases sometimes leave behind rich deposits of hematite. The metal iron is extracted from its red-brown crystals.

Steel screw made from iron

Galena
Galena is a sulfide of lead that sometimes comes in large shiny crystals. Around black smokers, it makes a powdery cloud of tiny crystals.

Lead, which comes from galena, has a low melting point, making it ideal for use in soldering wire

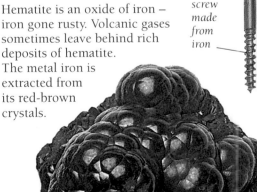

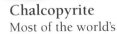

Peridotite
Peridotite is a heavy dark-colored rock, rarely found at the Earth's surface. The upper mantle may be made of peridotite.

Chalcopyrite
Most of the world's copper comes from the brassy-colored crystals of the mineral chalcopyrite. The chalcopyrite above is mixed with white quartz crystals.

SEDIMENTARY ROCKS

LAYER BY LAYER, sedimentary rocks are formed from material that previously made other rocks. This material is created by weathering, which breaks older rocks down into fragments and chemicals into solutions. Then, the grains and dissolved rock are transported by wind, rivers, and glaciers and eventually are deposited as layers, or beds, of sediment, along with any plant and animal remains trapped within. Over time, the layers are buried and squashed to become lithified, or hardened, into new rocks. Sedimentary rocks contain features that reveal their origins – the kind of rock that weathered into grains, the means of transport and deposition, and all the processes of lithification that turned loose sediment into hard rock.

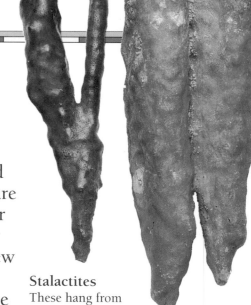

Stalactites
These hang from cave roofs when dripping water rich in dissolved limestone leaves a deposit behind.

Large, coarse pebbles   Medium-sized, coarse pebbles   Small, fine pebbles   Rock fragments   Quartz sand

GRAINS OF ROCK

Pebbles, gravel, and sand in a riverbed or on the beach are the raw materials for new sedimentary rock. As they tumble along in the water, pebbles and rock fragments may fracture by impact with each other, while continuing to be chemically weathered or dissolved. Where the grains settle, they may be buried by other layers. Water that percolates through the layers sometimes contains rock material in a solution. If this mateiral should come out of the solution to crystallize around the rock grains, the solution cements them together to make sedimentary rock.

Sandstone

The grains of sand that make up sandstone tell the rock's history. Those with a polished surface may be quartz grains that have been rolled around on a beach. Sand grains with a matt surface like ground glass show the sandstone was formed in a desert.

Breccia and conglomerate

A pebble beach might become hardened and lithified to make a conglomerate (below). The pebbles with sand grains between them are firmly held together by a rock cement. Breccia (right) forms in the same way, but its fragments are much rougher around the edges.

Rounded flint pebbles in this conglomerate have been smoothed by tumbling in water

Sharp rock fragments may pile up at the bottom of a cliff to form breccia

Layered limestones and volcanic ash rocks were used for the Colosseum in Rome, Italy.

Grindstone

This grindstone for grinding corn kernels was fashioned in Roman times from a hard conglomerate made of flint pebbles. The rough surface texture of the conglomerate is perfect for the job.

Bauxite

Bauxite is a mix of aluminum minerals left behind in tropical climates when all the other rock chemicals weather away.

The metal aluminum, used in foil, is extracted from bauxite

LIMESTONE

It is easy to see grains of sand on a beach that might one day become sandstone, but limestone's chemicals are transported invisibly – they are dissolved in water. Sea creatures and plants take carbon dioxide from seawater. This changes the chemical balance of the water, and as a result, the chemicals that make limestone – calcium and magnesium carbonate – separate from the water. They are deposited in thick layers of limy mud on the seabed to make limestone. Carbonates are the chemicals that make water hard.

Clay

Tiny clay grains weathered from other rocks travel suspended in water. These are what makes river water look muddy. When river water meets the sea, the clay flakes flock together into mud (left) that may form mudstone, claystone, or shale.

Cup and saucer made from clay

Clay brick

Shelly limestone

Sea creatures such as shellfish help make limestone by taking dissolved calcium carbonate from the water to make their shells. When the shellfish die, they sink into the limy ooze on the sea- or lakebed and help build up limestone.

Chalk

Chalk is a soft, pure limestone. Europe's chalk cliffs are made from the skeletons of tiny floating plants that lived in the sea more than 65 million years ago.

HOW COAL IS FORMED

Tree ferns and mosses that flourished in swamps of long ago sank into swampy water when they died. They were compressed to make peat. The deeper the peat was buried, the hotter and denser it became, until it was pressed into coal. The coal buried deepest became anthracite, the hardest, purest coal.

The plants did not rot, but formed a layer of peat. The remains of the plants are still visible.

Brown coal is peat that has been compressed by burial. Most of the water has been squeezed out. Some traces of the plants can still be seen.

Opal

Vividly colored opal coats cracks and cavities in sedimentary rocks. Sometimes opal transforms fossils trapped in rocks, replacing wood and shells and preserving their structures in opal.

Opal's colors make it an attractive gem

Sharp-edged flint arrowhead

Flint

Flint (above) is fine-grained silica, the same chemical as quartz. Flint breaks to make sharp edges, so it was ideal for making knives and arrowheads.

Colors refracted by layers of silica

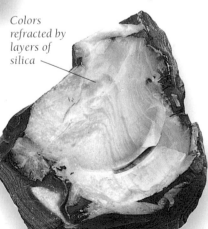

White opal

Black opal

Uncut opal

Sharp edges of gypsum crystals

The gypsum crystals in a desert rose grow in flakes that resemble rose petals

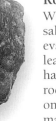

Black coal is hard and rich in carbon. It sometimes contains fossils of the plant matter that formed it.

Rock salt

When the sea or salty desert lakes evaporate, they leave salt, which hardens to become rock. Gypsum is one of the salts that makes up the rock.

Desert rose

When underground water in the desert evaporates, it leaves behind salts such as gypsum to crystallize. These gypsum crystals wrap around grains of sand.

Anthracite has been buried the deepest, and is the purest and cleanest coal to burn. It has a glossy surface, with few of the layers seen in black coal.

METAMORPHIC ROCKS

AS PLATES MOVE and crunch together, the rocks within are stretched, squeezed, heated – and changed. Metamorphic rocks are made from preexisting rocks, whether sedimentary, igneous, or metamorphic. When a rock metamorphoses, its minerals recrystallize and its original texture changes. Usually these changes happen deep inside the Earth's crust, where it is hot enough and there is enough pressure from the overlying rocks to make rocks recrystallize without melting. The recrystallization creates larger crystals and different minerals. At the same time, the rocks may become folded or crushed, so they get a new texture in which all the mineral grains are aligned according to the pressures on the rock. Several metamorphic rocks are featured on these two pages.

Quartzite
When a sandstone that is made entirely of quartz sand grains is metamorphosed, each grain of sand grows to a different shape in response to increased pressure. The once-rounded sand grains interlock and the spaces between are filled with quartz to form a tough new metamorphic rock, quartzite. It is much harder rock than marble.

Quartzite

LIMESTONE TO MARBLE
Intense heat changes limestone into marble, an even-grained, sugary-textured rock. Most limestones contain some chemicals other than calcium carbonate. These may be caught in the limestone as grains of sand, or wisps of clay. When the calcium carbonate recrystallizes to marble, it reacts with these other chemicals to make new metamorphic minerals, which come in many colors. The colored minerals may be in layers that become folded by the pressures of metamorphism. Folds in the green striped marble, for example, show that the original sedimentary layers get both thinned and thickened as they fold.

Limestone
Muddy gray limestone can be transformed into the multicolored marbles seen here.

Marble streaked with green

Gneiss
Heat and pressure change granite to gneiss. Gneiss shows dark wispy bands of mica curling around creamy white knots of feldspar.

Banded gneiss
The bands that stripe this rock show that directional pressure was high when it recrystallized.

Granite
This igneous rock is made of quartz, feldspar, and mica crystals. These are all more or less the same size, and are randomly scattered throughout the rock.

GRANITE BECOMES GNEISS
When granite is metamorphosed, the original crystals that make it up recrystallize. If metamorphism is not very intense, the new rock still looks like granite, but it takes on a new texture if there is also directional pressure, like the squeezing that happens in a mountain range. The new rock, gneiss, has a foliated texture. This means the minerals form wispy, more or less parallel bands.

White marble, made from pure calcium carbonate limestone, was used to build the Taj Mahal in Agra in northern India. This monument is decorated in fine patterns with colored slivers of marble and precious stones inlaid in the white.

Migmatite
This rock formed in heat so intense that parts of it melted. Migmatite shows no sign of the texture of the original rock, which might have been granite. Its banding has been intricately folded.

Only deep green beryl crystals like this can be called emeralds

Beryl has the same chemical composition as emerald

Ruby is colored a rich red by chromium chemicals

Sapphire is the same family as ruby, but is deep blue

Nephrite jade boulder from New Zealand formed in the high pressure of the Alpine Fault

Jadeite jade cut stones

Nephrite jade vase

Talc

Talc comes from the metamorphism of wet limestones containing sand. It is the softest mineral known, and has a smooth, slippery texture. Talc is the mineral used in making talcum powder.

Gemstones

Rocks are sometimes soaked with watery fluid during metamorphism. This helps recrystallization, so that bigger, clearer crystals grow. Many of the crystals shown here can be cut and polished as gemstones.

Jade

Jade minerals are tough, making them suitable for gemstones. They come from the metamorphism of dark igneous rocks and are often found in fault zones. There are two types: nephrite and jadeite.

Feldspar gives the marble its pale pink color

White marble (as shown in the statue above) has a grainy, regular texture that has been prized for carving since the time of the ancient Greeks.

SIR JAMES HALL BART.

Pink marble

The dark green pyroxene crystals in this Scottish marble were formed in a dolomite limestone.

White marble

White marble from southern Spain is recrystallized from a pure calcium carbonate limestone.

Black slate

Cubes of brassy-colored pyrite dot this rock.

Garnet schist

This schist grows garnet crystals in response to higher heat and pressure.

Grossular garnet

This garnet may be orange or green. Grossular garnets are found in Sri Lanka and Brazil.

Phyllite

The mica crystals within phyllite give it a shiny look.

MUDSTONE TO SCHIST

A dull, gray mudstone can be transformed by metamorphism into sparkling colored crystalline rocks. At different temperatures and pressures, different new minerals appear. The rocks here are shown in a sequence from left to right. The rocks on the left have been formed at the lowest temperatures and pressures, and those on the right, at the highest.

Garnets

Garnets are a group of minerals found in metamorphic rocks. Garnet forms crystals that come in many different colors. Their color depends on the chemistry of the original rock. Almandine, for example, gets its rich brownish-red color from iron. Garnet is a dense, hard mineral, tough enough to be used as an abrasive for grinding and polishing.

Almandine

Hessonite

Pyrope

Demantoid

Gray mudstone

This is a sedimentary rock formed in seas or lakes.

Kyanite schist

Its pale blue crystals formed in the depth and heat of a mountain range.

THE AGE OF THE EARTH

PLANET EARTH IS so old that it is hard to imagine its age: 4.5 billion years old. During the first 3.5 billion years of Earth history, the events that still shape its surface occurred: the first solid crust, the first life, the origins of the continents and the atmosphere, and the beginnings of plate tectonics. These events are recorded in the rocks that formed during this time. About 570 million years ago, there was a great explosion of life, and gradually the range of different plants and animals that now populate the Earth came into being. The remains of some of these life forms are captured in the rocks as fossils. These allow geologists to build up a picture of the Earth's long history.

Fossils provide valuable clues about the origin – and extinction – of life forms on Earth. In this 19th-century engraving, geologists are excavating fossil remains of a Mosasaurus from a cavern in Holland.

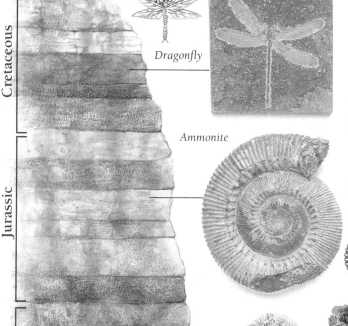

The buzz of insects
During the Cretaceous period, dragonflies and other winged insects buzzed in the skies and social insects such as butterflies appeared. They helped to pollinate flowering plants, which first evolved at that time. Flowering gave these plants a better chance of reproducing, despite the vast amounts of plant material munched by herbivorous (vegetarian) dinosaurs.

Dragonfly

Ammonite

The ammonite invasion
Ammonites were the predators of the Mesozoic seas. These free-swimming, carnivorous (meat-eating) creatures roamed far and wide. They became extinct at the end of the Cretaceous period, at the same time that the dinosaurs became extinct.

Reef builders
The shallow seas of the Mesozoic era formed when the continental shelves flooded as Pangaea split. In these sunlit waters, corals and bivalves (two-shelled mollusks) flourished, building vast limestone reefs.

Colonial coral

Vertebrates
Primitive fish some 400 million years ago were among the first vertebrates – animals with a vertebral (spinal) column. Later, on the continent of Pangaea, land vertebrates were to dominate, some feeding on the plants, which had become well established. Vertebrates also populated the seas and skies.

Ichthyosaur

Tree fern

Plants flourish on land
The first land plants appeared in Ordovician times. By the Carboniferous period 80 million years later, plants had become so diversified that vast forests flourished. Huge fernlike trees towered over steamy swamps. These forests later became preserved in the rocks as coal.

Afloat on the oceans
The oceans in Paleozoic times were populated with marine animals called graptolites. Many individual graptolites, such as these tweezer-shaped creatures, lived together in colonies, floating in the surface waters. Graptolites had already disappeared when a mass extinction at the end of the Paleozoic era wiped out several species.

Graptolite

HISTORY IN THE LAYERS
There is no place on Earth where rocks from every period throughout geological history lie piled one on top of another. In some places the rock layers represent big chunks of time. In Arizona's Grand Canyon, for example, the rocks span the whole Paleozoic era – 320 million years. But in most places, erosion has worn away millions of years of the Earth's history, and folding and faulting may have jumbled the layers that remain. To understand the whole sequence of time, geologists painstakingly match up one set of rocks with another, gradually building a stratigraphic (layered) column like the one on the far left.

The insects are about 40 million years old.

Trapped in amber
Not all fossils are found in rocks. These gnatlike insects are trapped in amber, a sticky resin that seeps from pine trees. The amber hardened, preserving their bodies.

Cretaceous

Jurassic

Triassic

Permian

Carboniferous

Devonian

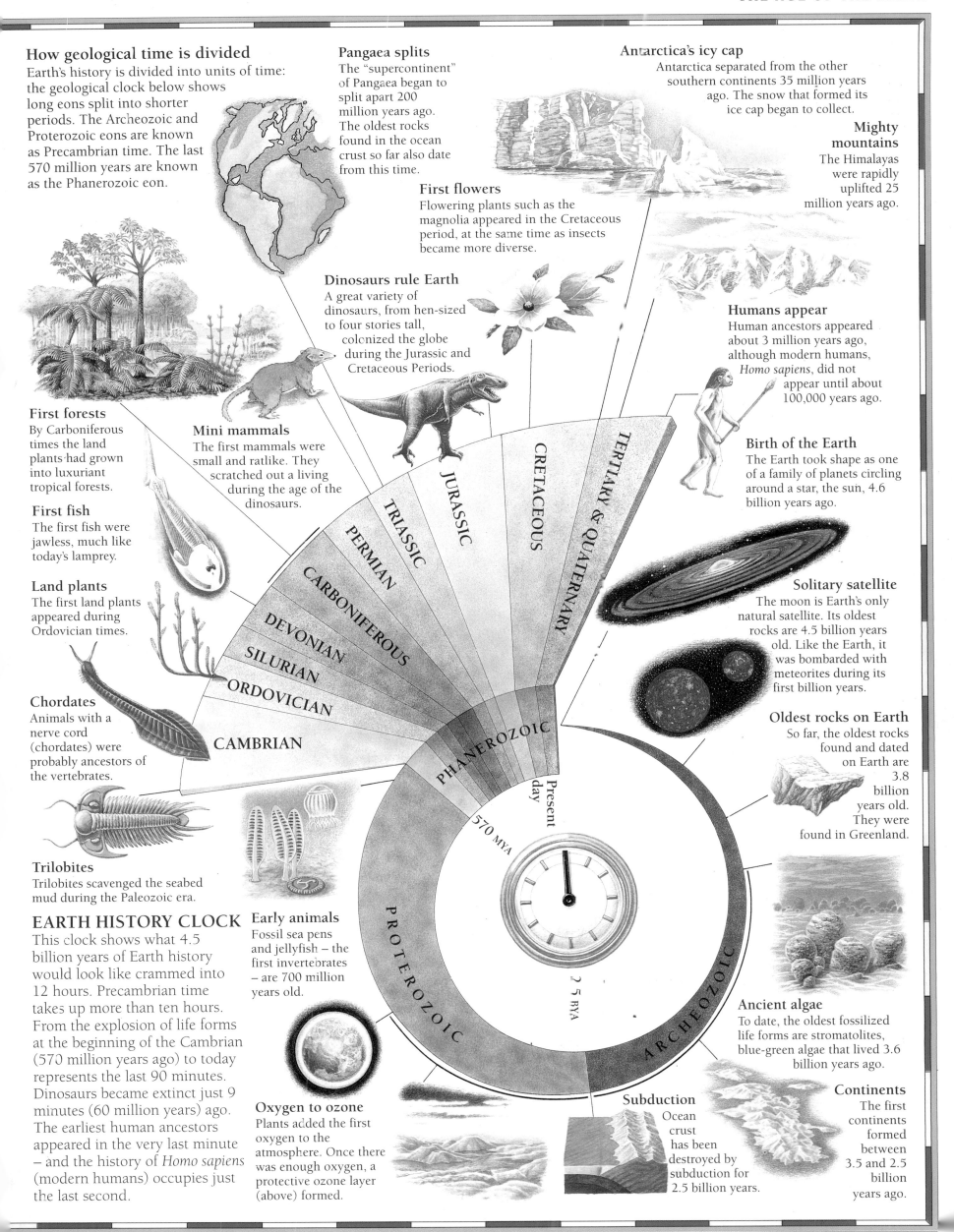

How geological time is divided
Earth's history is divided into units of time: the geological clock below shows long eons split into shorter periods. The Archeozoic and Proterozoic eons are known as Precambrian time. The last 570 million years are known as the Phanerozoic eon.

Pangaea splits
The "supercontinent" of Pangaea began to split apart 200 million years ago. The oldest rocks found in the ocean crust so far also date from this time.

Antarctica's icy cap
Antarctica separated from the other southern continents 35 million years ago. The snow that formed its ice cap began to collect.

Mighty mountains
The Himalayas were rapidly uplifted 25 million years ago.

First flowers
Flowering plants such as the magnolia appeared in the Cretaceous period, at the same time as insects became more diverse.

Dinosaurs rule Earth
A great variety of dinosaurs, from hen-sized to four stories tall, colonized the globe during the Jurassic and Cretaceous Periods.

Humans appear
Human ancestors appeared about 3 million years ago, although modern humans, *Homo sapiens*, did not appear until about 100,000 years ago.

First forests
By Carboniferous times the land plants had grown into luxuriant tropical forests.

Mini mammals
The first mammals were small and ratlike. They scratched out a living during the age of the dinosaurs.

Birth of the Earth
The Earth took shape as one of a family of planets circling around a star, the sun, 4.6 billion years ago.

First fish
The first fish were jawless, much like today's lamprey.

Land plants
The first land plants appeared during Ordovician times.

Solitary satellite
The moon is Earth's only natural satellite. Its oldest rocks are 4.5 billion years old. Like the Earth, it was bombarded with meteorites during its first billion years.

Chordates
Animals with a nerve cord (chordates) were probably ancestors of the vertebrates.

Oldest rocks on Earth
So far, the oldest rocks found and dated on Earth are 3.8 billion years old. They were found in Greenland.

Trilobites
Trilobites scavenged the seabed mud during the Paleozoic era.

EARTH HISTORY CLOCK
This clock shows what 4.5 billion years of Earth history would look like crammed into 12 hours. Precambrian time takes up more than ten hours. From the explosion of life forms at the beginning of the Cambrian (570 million years ago) to today represents the last 90 minutes. Dinosaurs became extinct just 9 minutes (60 million years) ago. The earliest human ancestors appeared in the very last minute – and the history of *Homo sapiens* (modern humans) occupies just the last second.

Early animals
Fossil sea pens and jellyfish – the first invertebrates – are 700 million years old.

Ancient algae
To date, the oldest fossilized life forms are stromatolites, blue-green algae that lived 3.6 billion years ago.

Oxygen to ozone
Plants added the first oxygen to the atmosphere. Once there was enough oxygen, a protective ozone layer (above) formed.

Subduction
Ocean crust has been destroyed by subduction for 2.5 billion years.

Continents
The first continents formed between 3.5 and 2.5 billion years ago.

Clock labels: JURASSIC, TRIASSIC, PERMIAN, CARBONIFEROUS, DEVONIAN, SILURIAN, ORDOVICIAN, CAMBRIAN, CRETACEOUS, TERTIARY & QUATERNARY, PHANEROZOIC, PROTEROZOIC, ARCHEOZOIC, Present day, 570 MYA, 2.5 BYA

INDEX

ACKNOWLEDGMENTS

Dorling Kindersley would like to thank Dorian Spencer Davies and Fran Jones for their work on the initial stages of this book, Sarah Cowley for final text fitting, David Gillingwater for design assistance, Louise Barratt for production work, James Mills-Hicks for cartographic advice, John Cope for text checking, and Jane Parker for the index. The author would like to acknowledge the help and advice of the following people and organizations: British Antarctic Survey; Simon Conway-Morris at Cambridge University; Rob Kemp at Royal Holloway, University of London; Martin Litherland at the British Geological Survey; Ian Mercer at the Gemmological Association; Ron Roberts; Robin Sanderson.

Additional illustrations Kuo Kang Chen, Fiona Bell Currie, and Andrew Robinson.
Picture research Joanna Thomas and Clive Webster.

Picture Credits
t=top, c=center, cb=center below, b=bottom, l=left, r=right

The publisher would like to thank the following for their kind permission to reproduce the photographs:

Dr Ian Boomer: 37tr; The Bridgeman Art Library/Royal Geographical Society, London: 5tc; John Cleare/Mountain Camera: 4bl, 26cl, 32c, 32bc; Bruce Coleman Ltd/Dr Charles Henneghien: 51cr/G. Ziester: 50cr; Steven J. Cooling: 33bl; Ecoscene/Farmar: 43tr; Mary Evans Picture Library: 10bl, 14tr, 21tr; Robert Harding Picture Library: 24cl; Dr Rob Kemp, Royal Holloway, University of London: 52bc; Frank Lane Picture Agency/USDA Forest Service: 5cr, 5br; Patricia Macdonald: 47br; The Natural History Museum, London: 8bl, 55bc; Natural History Photographic Agency/A.N.T.: 41tl; Peter Newark's Western Americana: 53tr; Oxford Scientific Films/Ake Lindau/Okapia: 48cl;

Ann Ronan at Image Select: 39tl, 62tr; Science Photo Library/Julian Baum: 5tr/Martin Bond: 52cl/Jean-Loup Charmet: 6bl/Tony Craddock: 30tr/Earth Satellite Corporation: 17tr, 23tr and cover/François Gohier: 8br/NASA: 23cl, 37cr, 41cr/Pekka Parviainen: 7cr/Alfred Pasieka: 62cr/Peter Ryan/Scripps: 39bl/Soames Summerhays: 19tl; John S. Shelton: 22c; Tony Stone Images: 49cr/Oliver Benn: 60cbr/Chris Haigh: 58c8br/Delphine Star: 46tr; Tony Waltham Geophotos: 4cr, 15tr, 27tr, 28cl; Zefa 47bl.

Additional photography on pages 56 to 63 by:
Andreas Einsiedel, Colin Keates and Harry Taylor of the Natural History Museum, and Tim Ridley.

Every effort has been made to trace the copyright holders and we apologize for any unintentional omissions. We would be pleased to insert the appropriate acknowledgment in any subsequent edition of this publication.

WITHDRAWN

SEAMAN MEMORIAL
PUBLIC LIBRARY
WEST BOYLSTON, MASS. 01583